BIOMEDICAL ENGINEERING

NEW YORK UNIVERSITY
MONOGRAPHS IN BIOMEDICAL ENGINEERING SERIES
GENERAL EDITOR: DANIEL J. SCHNECK

Biofluid Mechanics * 3
Daniel J. Schneck and Carol L. Lucas, Editors

Biomedical Engineering: Opening New Doors
Proceedings of the Fall 1990 Annual Meeting of the Biomedical
Engineering Society
Donald C. Mikulecky and Alexander M. Clarke, Editors

Biomedical Desk Reference
Daniel J. Schneck and Alan R. Tempkin

*Applications of Network Thermodynamics to Problems in Biomedical
Engineering*
Donald C. Mikulecky

Biomedical Engineering: Opening New Doors

Edited by

Donald C. Mikulecky
Virginia Commonwealth University
Medical College of Virginia
Richmond, Virginia

Alexander M. Clarke
Virginia Commonwealth University
Medical College of Virginia
Richmond, Virginia

NEW YORK UNIVERSITY PRESS
New York and London

Proceedings of the Fall Annual Meeting of
Biomedical Engineering Society,
Virginia Polytechnic Institute and State University

Library of Congress Cataloging-in-Publication Data
Biomedical Engineering Society. Fall Meeting (1990 : Virginia
Polytechnic Institute and State University)
Biomedical engineering : opening new doors : proceedings of the
Fall Annual Meeting of the Biomedical Engineering Society, Virginia
Polytechnic Institute and State University, October 21–24, 1990 /
editors, Donald C. Mikulecky, Alexander M. Clarke.
p.   cm.
ISBN 0-8147-7908-5 (cloth : alk. paper)
1. Biomedical engineering—Congresses.   I. Mikulecky, Donald C.,
1936–   II. Clarke, Alexander M., 1936–   .   III. Title.
[DNLM: 1. Biomedical Engineering—congresses.   QT 34 B6119b 1990]
R856.A2B575   1990
610′.28—dc20
DNLM/DLC
for Library of Congress   90-6566
                                                             CIP

TABLE OF CONTENTS

FOREWORD

This New York University Series entitled <u>Monographs in Biomedical Engineering</u> is intended to complement the very popular New York University textbook series in Biomedical Engineering. The Monographs series is dedicated to making available in a timely manner full Proceedings of Conferences, Meetings (such as the Annual Fall Meeting of the Biomedical Engineering Society), Workshops, Seminars, and other Specialty Symposia that have as their theme topics in the fast-growing field of Biomedical Engineering.

The Monographs series shall also publish single-authored works of a highly specialized nature, dealing with one specific area of Biomedical Engineering. Such works will have the character of being something between a long journal review article and a more comprehensive extensive textbook. Authors shall further be allowed to emerge from the anonymity of the scientific third-person passive in order to state their own (reasonably well supported) opinions about the subject matter being considered. Thus, this series intends for the personality of the author to be as evident as his or her work, and originality and creativity are strongly encouraged.

Books in the Monographs series are offset from author-prepared, camera-ready copy; which allows quick turnaround, makes the work both timely and current, and insures affordability of the published product. Potential authors (or professional organizations) wishing to contribute to this series should contact the general editor, Dr. Daniel J. Schneck, at Virginia Polytechnic Institute and State University, Department of Engineering Science and Mechanics, 227 Norris Hall, M/S ESM 0219, Blacksburg, Virginia 24061. For Conference Proceedings, please include a preliminary program and other relevant information. For single-authored monographs, include a proposed table of contents, rationale for the work, and some samples of published work. For edited works, include as well a list of potential authors, with titles for their contributions.

DANIEL J. SCHNECK
Virginia Tech

PREFACE

Biomedical Engineering: Opening New Doors is a report of the proceedings of the 1990 Fall Annual Meeting of the Biomedical Engineering Society. This volume contains five sections, each devoted to a special area of biomedical engineering. The spectrum spanned by these presentations indeed is an opening of many new doors into the research of the next century. Some of these new areas have already become easily recognized as topics for the popular press, while others remain less conspicuous as their students push forward without as much fanfare. Nonetheless, each in its own way stakes out territory for future exploration and new sources of excitement.

New methods of mathematical modeling, imaging and computation encompass neural networks and other forms of artificial intelligence. Networks also appear as network thermodynamics is utilized. Artificial intelligence is applied to life support and other applications. Many techniques for looking at organs and tissue designed to perturb these structures as little as possible are explored. The new uses of biotechnology are also discussed.

All together these diverse topics have a common theme, namely the opening of doors to the future of biomedical engineering. We sincerely hope the reader will find them as exciting as we have while preparing this volume. We want to give special thanks to the session organizers, Dr. Alex M. Clarke (Physiological measurements and life support technologies), Dr. John Tyson (Mathematical Models in Molecular Biology and Physiology), Dr. Arthur Johnson (Biotechnology), Dr. Michael Merickel (Non-invasive detection of tissue and organ failure: Imaging), and Dr. Gene A. Tagliarani (Engineering aspects of cognitive sciences) We also wish to thank the Whittaker Foundation for their generous support. A special thanks goes to Jack Lilly, Program Development Specialist, and his staff at the Donaldson Brown Center for Continuing Education, for their gracious hosting of this conference, and to

Jason Renker and all of his colleagues at the N.Y.U. Press for their encouragement and timely publication of the Proceedings volumes. We are proud to be part of this great team and are grateful to so many dedicated people for helping make this conference a success.

Donald C. Mikulecky
Editor

Alex M. Clarke
Editor and Session Organizer   and C   h   a   i   r   , Physiological Measurements and Life Support

October, 1990
Virginia Commonwealth University,
Richmond, Virginia

I.

PHYSIOLOGICAL MEASUREMENTS
AND LIFE-SUPPORT TECHNOLOGIES

A PROMPTING AND DATA COLLECTION SYSTEM FOR USE IN MEDICAL

DEPARTMENTS DURING CPR

N. Patel, P.L. Thery, A.M.L. Conner, J.P. Ornato, and

A.M. Clarke, Medical College of Virginia, Virginia

Commonwealth University, Richmond, VA, 23298-0694.

ABSTRACT

The development of systems for accurate recording of
both the medical protocols and treatment undertaken, together
with the resulting electrocardiographic, blood pressure, and
other physiologic responses, all carefully "time-stamped,"
was undertaken to help optimize cardiopulmonary resuscitation, and provide the detailed records necessary for continued treatment while the patient is in the hospital. These
systems also allow scrutiny of the procedure for accuracy,
billing, and even, should a Regional or National Consortium
be instituted, rapid determination whether changes in the
recommendations of the ACLS and other groups result in increased survival rates. In this paper, the parameters of the
systems will be described. The potential of the system lies
in the adoption of a national standard for the reporting of
this data, such that future changes in treatment protocols to
improve CPR can be taken to trial more efficiently

INTRODUCTION

In the pressured, highly charged atmosphere surrounding
a cardiac resuscitation attempt in the emergency department,
meticulous documentation of all treatment is essential for a
hospital to provide optimal health care. Two microcomputer-
based data management and acquisition systems have been developed. The systems have as a central requirement that they
be user-friendly and interactive, and that they be easily and
intuitively used by physicians and nurses during a cardiac
emergency. The system is a more accurate, efficient, and

1

reliable method of data management than the hand-written records currently compiled at most hospitals.

The modern microcomputer has numerous intrinsic characteristics lending itself to an excellent emergency care data management system. The speed of the computer allows quick and simple entry of important events in much greater detail than hand-written efforts. These inputs may be easily time-stamped and stored in a chronological manner by the computer for real-time and subsequent review. The electrocardiogram and other analog waveforms monitored during the cardiac emergency can easily be sampled by the computer and stored for later reconstruction and analysis.

While microcomputer-based documentation systems for the intensive care unit of the hospital are of great value in their setting, cardiac emergencies present quite different obstacles to be overcome by a data management system. An ICU system is usually designed to accept relatively infrequent entry of information concerning numerous patients over a relatively long period of time (days to weeks). In contrast, a cardiac resuscitation data management system is required to frequently collect in-depth information concerning one patient over a short period of time (minutes to hours). During the cardiac resuscitation, the system's emphasis must be speed, ease of entry, and ease of review. These considerations are important, but not critical, in the hospital ICU setting.

The set of entries to be logged by a data management system during a cardiac resuscitation is unique, including selections such as defibrillation, detection of various arrythmias, and administration of drugs commonly used during CPR. An important need of the cardiac resuscitation data management system is a combination of in-depth logging of treatment events and digital storage of monitored patient signals. Most ICU systems which allow detailed entry of clinical events do not include a data acquisition module and are not designed for fast-paced entry.

THE SYSTEMS

Two systems have been developed at the Medical College of Virginia in conjunction with the Hospitals Emergency Department. The first is written in "C", requires a 80286 or 80386 based microcomputer compatible with the Microsoft disk

operating system, 512 kilobytes of RAM, and a 20 megabyte or
larger hardfile that has an access time of 30 ms or better.
The data acquisition board used for this particular system is
the Labmaster DMA <tm> produced by Scientific Solutions, with
12 bits of analog to digital resolution, and 40 kHz sampling
capability. The program is written in modules, simplifying
changes in both the the hardware and software configurations,
as well as facilitating changes in the treatment protocols.

     The second system is written in Borland Turbo Prolog,
chosen for its increased in graphic capability and the potential for the development of an Artificial Intelligence shell.
Several modules have been written in Assembly Language, where
Prolog was too slow. Use of Extended Memory as well as 80386
and 80387 specific code limits this system to a 80386/80387
based machine, preferably with 2 megabytes of RAM or more.
The data acquisition board is the Metrabyte Dash 16 <tm>.
Interface to the patient is made through the patient monitor
which has is the standard at the MCV Hospitals, the Mennen
Horizon 2000.

     Upon activation of the system, a menu, a protocol flow
sheet, and pull-down window appears, with the most prevalent
condition observed at the admission of a patient to the
resuscitation unit of the Emergency Department displayed.
Movement through the cascaded menus is possible with either a
"Mouse," or the cursor keys. Movement has been designed to
be as intuitive as possible, with conditions and responses
most often encountered being in the default position in the
individual menu. Data collection to the flowsheet commences
with activation of the system. However, the analog signal
acquisition is activated only on command, as the connection
to the patient is sometimes delayed during the first few
moments of the CPR attempt. As many as 5 analog signals may
be displayed in a window occupying the bottom third of the
screen. For the Mennen monitor, the digital signal containing the mean, diastolic, and systolic blood pressures, heart
and respiration is read, displayed, and automatically entered
when vital signs are requested.

     Included in the documentation are drug administrations,
heart rhythm observation, vital signs, intubations, and
numerous other cardiac resuscitation procedures. Upon selection of a particular item, the data management system leads
the user through a series of cascading menus prompting entry
of the most pertinent treatment information. The change of
menus, cursor movement, and display update appear instanta-

neous to the user. The system provides recommendations for drug dosages based upon American Heart Association standards. Instructions for the mixing of infusions are provided on the computer display and may be directed to a printer for use by the medical staff.

The analog channel display and flowsheet of logged events are located on the computer display for immediate review by the emergency team. Storage of six second "snap-shots" of six sampled analog waveforms is performed upon user request and automatically after the logging of certain medical procedures. The execution of the storage takes a maximum time of one second and does not impede the real-time signal display or the logging of events. At the conclusion of the emergency, these waveforms may be reconstructed and directed to a printer for inclusion in the patient file.

The system optimizes patient care by providing a quick and reliable method to record and display detailed patient treatment during the emergency. At the conclusion of the emergency, detailed information is available to the hospital for analysis and review. The final patient record is comprised of the documented events of the emergency, patient demographic information, and reconstructed patient waveform "snapshots". A separate binary file is available for subsequent statistical analysis and waveform reconstruction.

SUMMARY

The system provides the speed, accuracy, and reliability not currently available to hospital cardiac resuscitation units. The system is unique in that it provides in-depth data entry of emergency events, along with display and storage of digitized patient vital signals. Implementation of the system will improve patient care as well as evaluation of cardiac emergency procedures and programs. Most important, standardization of a method for the collection and reporting of this form of data would allow the building of a national database. Changes in the American Heart Association protocols could be quickly evaluated, and the exchange of information on CPR among major medical facilities would be markedly enhanced as would the updating of the protocols used in smaller and rural hospital emergency treatment centers

Supported by the Foot and Levy Fund, and the Asmund S. Laerdal Foundation, Inc.

DESIGN FOR A 512 CHANNEL CARDIAC MAPPING SYSTEM

Patrick D. Wolf, Dennis L. Rollins, Timothy F. Blitchington, Raymond E. Ideker, William M. Smith

Departments of Medicine and Pathology
Box 3140 Duke University Medical Center
Durham, NC 27710

ABSTRACT

We have designed and are constructing a 512 channel cardiac mapping system for our research laboratory. This new system will be capable of recording 512 channels of data at a sampling rate of 2 kHz per channel. The dynamic range of each channel will extend from less than $1\mu V$ to $\pm600V$. The data throughput rate of 2 megabytes per second will be recorded continuously on VHS cassettes. A real time hardcopy display of 32 channels will be generated and a CRT will be used to display the system status and to present detailed displays of specific events. The system will provide a buffered interface for transferring data to our laboratory computer network.

The design is being implemented with the goal of creating a flexible system which can adapt to new products. This goal has been attained by designing the system hardware around public domain standards wherever possible and writing the software using a real time multi-tasking system. Writing each functional unit as a separate task will minimize the software interaction between mapping system functions. Creating a flexible design will allow us to incorporate new storage, processing and display technologies as they become available with a minimal effect on the total system.

INTRODUCTION

In our laboratory we study the basic mechanisms of ventricular fibrillation, and other forms of arrhythmias and the effects of therapeutic intervention with externally applied electric fields. The basic tool for this research is a cardiac mapping system. The mapping therapeutic intervention with externally applied electric fields. The basic tool for this research is a cardiac mapping system. The mapping system is basically a large scale data acquisition system. Electrograms from hundreds of electrodes on the heart are amplified, filtered, converted to digital form and then stored and processed. In our laboratory setting, the mapping system is used by many different investigators. In order to have the data analysis for multiple studies go on simultaneously, we separated the analysis software from the acquisition hardware and use a laboratory network of workstations to analyze the data [1]. In this paper we will present the design which we are currently implementing for the data acquisition portion of the system.

There are four major functions of the data acquisition hardware. First, the analog signals must be processed in the analog domain and converted to digital data. Second, the data must be stored continuously on a removable medium. Third, there must be the capability to display the signals on a screen and on paper so the investigator can have real time feedback during the experiment. Finally, the system must be capable of interfacing to our laboratory analysis system. The design for the implementation of these functions is discussed below.

We are designing our fourth generation mapping system. Our previous systems have relied on custom built hardware to achieve the functionality that we required. These systems lacked the ability to change incrementally as newer technologies became available. The hardware and software for the acquisition, display and storage of analog waveforms could be changed only with a redesign. A major goal of the current project was to build a system which could incorporate new devices and new software without a major system overhaul. We have met this goal by using hardware and software standards wherever possible, and by using a multi-tasking design for implementing the software. Incorporation of these features in the system will allow the integration of new devices and software with a minimum effect on existing system components. A second major difference between this design and our other systems is the real time display capability. This design includes two programmable real time display devices. The increased availability of graphics processors has allowed us to create a more useful display for experiments.

Our investigations have shown that it is possible to create reentrant tachycardias in the thin wall of the right ventricle in a space smaller than 4 square centimeters [2]. Adequate recording of these events required electrode spacing of two millimeters. To obtain this level of resolution over larger areas of the heart, we will require more channels than the 128 we can currently record. This mapping system will allow us to record from 512 channels simultaneously. Cabo showed that for unipolar electrograms, it is easier to distinguish distant from local electrical activity when analyzing waveforms which include information in the 700 to 800 Hz range [3]. For this reason, we have increased the minimum sampling rate of the system from 1 kHz to 2 kHz. The system will also be capable of increased sampling rates on fewer channels: 4 kHz sampling on 256 channels, 8 kHz sampling on 128 channels, and 16 kHz sampling on 64 channels.

DESCRIPTION

An overall system diagram is shown in figure 1. The system has two basic units, the analog processing and conversion unit and the storage and display unit. The storage and display unit consists of four functional modules outlined in figure 1: (1) a data link from the first unit, (2) a display module, (3) a storage module, and (4) a buffered link to our laboratory computer. These modules will all be built into the same chassis and will communicate over the VMEbus. This bus standard (IEEE 1014) was chosen because of the availability of products from multiple vendors, and because it provided enough bandwidth to implement the design with some room for future expansion. Following is a more detailed description of the acquisition unit and the four modules in the display and storage unit.

Analog Signal Processing

To study defibrillation we need to be able to record signals over a very broad dynamic range. A signal resolution of microvolts is required for recording electrograms generated by the myocardium; for measuring the voltages in the heart generated by defibrillation shocks, a signal range up to hundreds of volts is needed. We will use an attenuation circuit combined with programmable gain amplifiers to cover this large dynamic range. The attenuator circuit

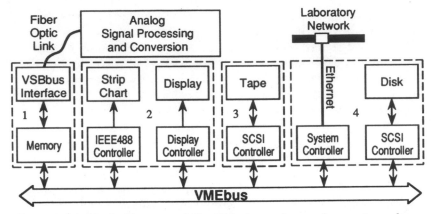

Figure 1. A block diagram of the 512 channel mapping system showing the relationship between the functional units. Dashed lines outline the modules described in the text.

will consist of a 200:1 voltage divider built using a 100M Ohm resistor network. The network will be switched into the signal path using electromechanical relays [4]. This relay switching circuit will also be used to switch between recording differentially on each channel to recording in a single ended mode with the inverting input for each amplifier derived from a fixed reference.

After passing through the attenuator circuits the signal will be processed by the circuit shown in figure 2. Amplifiers A1-A3 and resistors R1 and R2 make up a switchable gain instrumentation amplifier. With switch S4 open, the gain will be unity; with it closed the gain will be eight volts/volt. Following this stage will be a unity gain, 2 pole, smoothest response, low pass filter with four programmable cutoff points. This is the antialiasing filter for the four sampling rates, with 3 dB points set at 500 Hz, 1 kHz, 2 kHz, and 4 kHz. To limit the bandwidth at the low end of the spectrum, we use the variable high pass filter made from C1, S2 and A5 with settings of DC, .05, .5, and 5 Hz. Single ended gains of 1, 10, and 100 will be implemented by the variable gain amplifier made from A6 and S3. This stage will be followed by a sample and hold amplifier circuit.

All the components for the circuit shown in figure 2 with the exception of 5 discrete components will be mounted on a thick film

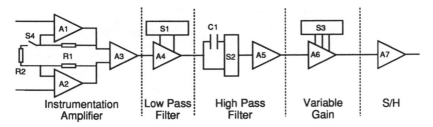

Figure 2. Schematic diagram of the components for a single analog channel.

hybrid circuit measuring approximately 1.75 by 1.0 inches. Also contained on the hybrid will be a latch for holding the 8 digital signals used to control the analog circuit characteristics.

Thirty-two hybrid circuits, a multiplexer and a 14 bit A/D converter will be placed on a circuit board along with the components to load the channel latches from a controlling microprocessor. Sixteen circuit boards will be combined to form the 512 channel unit. The microprocessor used to control the front end will format and add to the data stream eight header words each millisecond. This data will describe the characteristics of the front end channels as well as provide additional user information [5]. The data words from the 16 converters and the computer will be bussed to a common controller which will serialize the data and send it over a fiber optic link to the display and storage hardware. The data rate on this fiber will be approximately 20 megabits per second.

A 68020 microprocessor with memory, serial lines and a subsystem bus interface will be used to control the characteristics of the amplifiers, to format the header data to be entered in the data stream, to communicate with the user, to implement automatic gain control and other signal integrity software, and to provide real time control of the amplifier characteristics. The amplifiers with the controlling microprocessor will form the stand alone data acquisition unit with isolated serial communication to the remainder of the system.

At the receiving end of the data fiber will be a hardware module for converting the bit stream back to parallel data and placing it in memory on the VMEbus. The memory buffer will be dual ported with access from either the VME or the VME Subsystem Bus (VSBbus). The fiber link will dump the data into memory using the VSBbus port. All the remaining devices will access the data through the VMEbus port. This design eliminated from the VMEbus transaction load the 2 megabytes per second of bus bandwidth needed just to acquire the data. Once in the VMEbus memory, the data will be available to the other devices in the system.

Data Storage

We consider continuous data storage an important feature of our mapping system designs. In keeping with this philosophy, we have designed the system to record the 2 megabyte per second data stream on a Honeywell VLDS tape recorder (Honeywell Test Instruments Division, Boulder, CO). The VLDS is a helical scanning tape device which can record up to 4 megabytes per second on VHS cassettes. Using a two channel VLDS we will be able to record our data stream continuously for about 40 minutes. Each VHS tape will hold 5.2 gigabytes of data. The VLDS will be interfaced to the VMEbus using a Small Computer Systems Interface (SCSI, ANSI X3.131) high speed host bus adapter capable of synchronous transfers at 4 Mb per second.

Real Time Display

Two types of real time display are required in our system. First, a visual display for communicating with the users allowing them to setup and control the system according to their requirements, for making detailed examination of single channels or events, and possibly for displaying results and maps. This device will be a medium resolution (1280x1024) color monitor driven by a VMEbus graphics controller card. The controller will be programmed by a 68030 based single board computer with floating point processor and high speed direct memory access (DMA) capability.

The second type of display which is required is a strip chart recorder. This device provides a quick mechanism for displaying and recording events and presents a hardcopy of the signals to the investigator in a familiar format. Because of its digital capability, we will use the Astro-Med MT9500 strip chart recorder (Astro-Med Inc, West Warwick, RI). This device can be programmed to display 32 channels across the page allowing the user to verify the function of many channels at a time. The recorder's digital interface is through the IEEE 488 instrumentation bus. The VMEbus to IEEE 488 bus interface will be handled by a 68000 based single board computer acting as a controller.

Laboratory System Interface

There are two modes used to analyze the mapping system data. The first mode can be considered on-line, where the next step in an experimental or surgical protocol depends on the analysis of the data obtained from the previous step. The second is off-line analysis where the data from a particular procedure is analysed after the experiment is over. On-line analysis requires that the mapping system be capable of transferring data to the laboratory computer system while continuous recording is in progress. The second mode requires only that data be transferred from the tape to the network. Our laboratory computer system is a network of over 20 computers tied together with Ethernet. All of the machines are running the UNIX† operating system and none of the machines has enough memory to handle a 2 Megabyte per second data stream for the 15 to 30 seconds needed to acquire data directly from the mapping system. Additionally, the standard disks available today cannot store this much data at this rate in a UNIX file system. In order to get the mapping system data onto the laboratory network requires a temporary storage mechanism on the mapping system and a mechanism for transferring the data at a slower rate. To solve the first problem we are using a large, fast disk connected to the VMEbus using a SCSI bus host adapter. The disk is fast enough to store the data in real time in continuous files. This disk will be used as a large buffer for storing multiple runs during off-line analysis, and for buffering the transfer during on-line analysis. The mapping system will actually transfer the data from the disk buffer to the laboratory network over Ethernet. File transfers will be made using the standard TCP/IP communications protocol. With this design, the mapping system will appear as another node on the laboratory network.

---

† UNIX is a trademark of AT&T Bell Laboratories.

Software

All the software for the system will be written in C under the VxWorks‡ real time operating system This operating system will allow us to develop the code for the mapping system processors on the laboratory network of UNIX workstations and contains the facilities for downloading and debugging the software over the network. In addition, VxWorks contains the software for implementing a multi-tasking design and for communications between UNIX and VxWorks processors over Ethernet.

The software for the system is being designed using tasking concepts, where each of the mapping system functions will be programmed as a separate module using the facilities of semaphores and shared memory to communicate between tasks. This software design will allow us to implement changes in one task with a minimum effect on the software from other tasks. Four single board computers have been designed into the system to implement the various functions. The first processor will control the data acquisition hardware; the second processor will run the tasks which control the fiber optic cable to memory link, the tape recorder and the disk; the third will run the task to control the IEEE 488 link to the strip chart recorder; and the last board will manage the Ethernet link and the graphics processor tasks.

CONCLUSION

We have described the design for a 512 channel mapping system. By designing the system in modular form using hardware and software standards wherever possible we hope to be able to take advantage of new products and to reprogram the functions of existing modules with a minimum of interaction with the other devices. We should be able to take advantage of new storage technologies as they become available for the SCSI bus, also, new graphics processors designed for the VMEbus can be incorporated as they become available. Finally, as higher speed processing devices are developed and mapping techniques become further refined, real time analysis of electrograms with near real time mapping is a distinct possibility with this design.

---

‡ VxWorks is a trademark of Wind River Systems, Inc. Emeryville CA.

ACKNOWLEDGEMENTS

This work was supported in part by National Institutes of Health Research Grants HL-28429, HL-33637, HL-42760, HL-44066 and by National Science Foundation Engineering Research Center Grant CDR-8622201.

REFERENCES

[1]   N. Danieley, P.D. Wolf, R.E. Ideker, and W.M. Smith, "A workstation network for cardiac electrophysiology," in *Proc. Computers in Cardiology*, 1988, pp. 469-472.

[2]   D.W. Frazier, P.D. Wolf, J.M. Wharton, A.S.L. Tang, W.M. Smith, and R.E. Ideker, "Stimulus-induced critical point: Mechanism for the electrical initiation of reentry in normal canine myocardium," *J. Clin. Invest.*, vol. 83, pp. 1039-1052, 1989.

[3]   C. Cabo, J.M. Wharton, P.D. Wolf, R.E. Ideker, and W.M. Smith, "Activation in unipolar cardiac electrograms: A frequency analysis," *IEEE Trans. Biomed. Eng.*, vol. 37, pp. 500-508, 1990.

[4]   P.D. Wolf, J.M. Wharton, C.D. Wilkinson, W.M. Smith, and R.E. Ideker, "A method of measuring cardiac defibrillation potentials," in *Proc. 39th Annual Conf. on Engineering in Medicine and Biology*, 1986, p. 4.

[5]   P.D. Wolf, R.E. Ideker, and W.M. Smith, "A data stream formatter for a cardiac mapping system," in *Proc. 8th Annual Conf. of the IEEE Engineering in Medicine and Biology Society*, 1986, pp. 491-493.

Nonlinear Techniques for Characterizing Heart Rate and Blood Pressure Variability

Daniel T. Kaplan
Colin Research America
One Kendall Square, Suite 2200
Cambridge, Mass. 02139

Monitoring of heart rate and blood pressure has typically involved making occasional measurements or averaging closely spaced measurements. The goal of monitoring has been principally to detect instances where heart rate or blood pressure goes "out of bounds." Variation in heart rate at frequencies greater than tenths of hertz has been considered noise and ignored or averaged out.[1] Until recently, blood pressure could be measured continuously only with the use of invasive arterial lines; in a non-invasive setting, measurements could be made at best on the order of a minute apart.

Interest in variability in heart rate and blood pressure has become well established only in the past decade. The techniques used include a simple calculation of variance, parametric and non-parametric descriptions of the power spectrum and transfer functions, estimation of complete models of cardiovascular regulation, and, more recently, calculation of dimension and entropy. Work in this field indicates that variability in heart rate and blood pressure can be used to measure the tone of the sympathetic and parasympathetic branches of the autonomic nervous system, and has pointed out several areas in which monitoring of heart rate variability may

---

[1] A counter-example is respiratory sinus arrhythmia, although this is infrequently used for diagnostic purposes.

have clinical utility.[2] Recently, non-invasive means of measuring blood pressure on a continuous basis have been developed, making it much easier to investigate variability in blood pressure as well as heart rate.[3]

Linearity vs. Nonlinearity

The most widespread type of analysis of heart rate and blood pressure variability focusses on *linear* systems. The fundamental assumption is that the important characteristics of heart rate and blood pressure variability are the poles and zeros in the power spectra of the signals. The apparent irregularity in heart rate and blood pressure is ascribed to exogenous noise, which is shaped and colored by the resonances and nulls of the cardiovascular control system. From the linear perspective, the shaping and coloring are the phenomena of physiological interest; the noise is not.

An alternative point of view is that irregularity and "noise" are an essential part of the cardiovascular system, and the the irregularity may be — at least in part — the product of the internal workings of the system and is not due to exogenous influences. This point of view is largely motivated by the theory of *chaos*, which has shown that simple, deterministic systems can generate quite irregular- and noisy-looking outputs without any exogenous influences. Such systems are always *nonlinear*.

Reconstructing a System from a Measured Signal

The techniques that most usefully characterize nonlinear, chaotic systems are different from those used to study linear systems. In general, the analysis of signals in the context of nonlinear dynamical systems is based on attempts to reconstruct deterministic dynamics that can reproduce the signal in question. The most useful tool in this regard is the technique for *embedding* the signal.

Deterministic dynamical systems can be written in the form of differential or difference equations, e.g.:

[2]e.g., D. Gordon *et al.*, "Heart-rate spectral analysis: a noninvasive probe of cardiovascular regulation in critically ill children with heart disease," *Pediatric Cardiology* 9:69-77, 1988; R Kitney *et al.*, "Heart rate variability in the assessment of autonomic diabetic neuropathy" *Automedica* 4:155-167, 1982; Pagani *et al.*, "Power spectral analysis of heart rate and arterial pressure variabilities as a marker of sympatho-vagal interaction in man and conscious dog" *Circ. Res.* 59:178-193, 1986

[3]J. Eckerle and H. Lippincott, "Advances in arterial tonometry during the last decade" *Proc. AAMI 25th Annual Meeting*, 5-9 May, 1990

$$\dot{x}_1 = f_1(x_1, ..., x_n) \tag{1}$$
$$\dot{x}_2 = f_2(x_1, ..., x_n)$$
$$\vdots$$
$$\dot{x}_n = f_n(x_1, ..., x_n)$$

For nonlinear dynamical systems, the functions $f_i()$ are, of course, nonlinear. The vector $\vec{x} \equiv (x_1, ..., x_n)$ is the *state* of the system. The functions $f_i()$ describe a dynamical rule for how the state changes in time. Measurements made on the system are

$$z(t) = g(\vec{x}(t)) \tag{2}$$

A remarkable fact is that the state $\vec{x}(t)$ of the original system can be represented by a vector constructed from the *scalar* time series $z(t)$ as

$$\vec{z}(t) \equiv (z(t), z(t - \tau), z(t - 2\tau), ..., z(t - (m-1)\tau)) \tag{3}$$

Here, m is called the *embedding dimension* and τ is called the *embedding lag*.

The first step in analyzing a signal from a nonlinear dynamical system is to embed the signal as in equation 3. The only theoretical guidance in choosing m is that $m \geq 2n + 1$. However, since one does not generally know n in advance, it is typical to repeat the analysis at increasing m until the results no longer change. The choice of τ is more controversial: τ should be something like the delay at which the signal's autocorrelation function falls to $\frac{1}{e}$.

Analyzing the Reconstructed Signal

Once the measured signal has been embedded, several forms of analysis are possible. Perhaps the most conceptually clear is to use the vector time series $\vec{z}(t)$ to construct a model of the system's dynamics. This can be done by assuming that the dynamics in the m-dimensional space are piecewise linear.[4]

The most commonly used forms of analysis involve the *correlation integral*, $C_m(r)$, which is simply the number of pairs of points in the $\vec{z}(t)$

---

[4]G. Sugihara and R. May, "Nonlinear forecasting as a way of distinguishing chaos from measurement error in time series" *Nature* **344**:734-741

time series that are closer than distance r to one another.[5] The $\vec{z}(t)$ time series carves out a *trajectory* in the m dimensional embedding space. This trajectory may be characterized by a *dimension* which will in general be less than the embedding dimension. (For chaotic systems, the dimension will be fractional.) The dimension of the trajectory, ν, can be calculated from $C_m(r)$ as

$$\nu(r) \equiv \frac{d \ln C_m(r)}{d \ln r} \tag{4}$$

Note that I have written ν as a function of r. An important issue is the choice of r at which to evaluate ν. I will address this issue below.

Another way of characterizing nonlinear dynamical systems has to do with their predictability. This can be quantified by the *entropy*, which also can be calculated from $C_m(r)$.

Chaos and Complexity

A number of workers have studied heart rate using variants of the methods outlined above.[6] Their goal has been to determine whether heart rate dynamics are deterministic chaos, i.e., whether despite heart rate's apparent irregularity, it arises from a deterministic dynamical system. The results of the analysis are somewhat difficult to interpret. Usually the "control" hypothesis is that the heart rate is white noise; it is not difficult to show with the above techniques that a heart rate signal is different from white noise. This does not necessarily mean that heart rate comes from a deterministic chaotic system, however.

The important problem for monitoring heart rate and blood pressure variability is to be able to quantify variability with a small set of numbers so that clinical correlates can be found. Even if heart rate and blood pressure are not chaotic, nonlinear techniques such as calculation of dimension and entropy may be useful statistics for quantifying variability. These statistics measure something that might be called "complexity."

When using dimension and entropy calculations to measure complexity, one needs to take a slightly different approach than that used when looking

---

[5] P Grassberger and I Procaccia, "Measuring the strangeness of strange attractors", *Physica* **9D**:189, 1983; G. Mayer-Kress, ed., *Dimensions and Entropies in Chaotic Systems*, Vol. 32 of *Springer Series in Synergetics* (Springer-Verlag, New York, 1986)

[6] A Babloyantz and A Destexhe, "Is the normal heart a periodic oscillator?" *Biol. Cybernetics* **58**:203-211, 1988; G. Mayer-Kress *et al.*, "Dimensional analysis of nonlinear oscillations in brain, heart and muscle" *Mathematical Biosciences* **90**:155-182, 1988

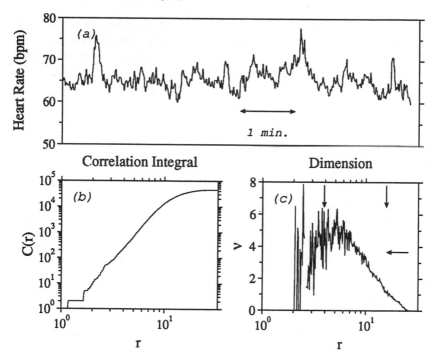

(a) A time series of heart rate from a 77-year old human subject.

(b) The correlation integral, $C(r)$, for the data shown calculated with an embedding dimension of 10 and an embedding lag of 3.5 sccs. Note that the graph is plotted on log-log scales.

(c) The dimension, $\nu(r)$ as calculated from the graph in (b) using equation 4. The vertical arrows indicate the length scales over which the average ν is calculated; the value of this average is shown on the horizontal arrow.

for chaos. Figure 1 shows an example of a heart rate time series from a 77-year old human subject. The hallmark of chaos is straight line segments in the graph of $\ln C(r)$ vs. $\ln r$. Techniques for looking for chaos usually search for length scales of r at which the ν in equation 4 is constant.' From the graph of $\nu(r)$ in Figure 1, one can see that ν is approximately constant at $r \approx 5$ — the peak of the $\nu(r)$ curve. One might be tempted, then, to assign to this data segment a dimension of approximately 5. This would be seriously misleading.

Since the signal shown in Figure 1 has been embedded with $m = 10$, variations in the signal at the shortest length scales ($r \approx 2$) correspond to heart rate changes of approximately 0.2 beats per minute. This is within the quantization noise for heart rate derived from an ECG sampled at 250hz. Thus, the dynamics at this short length scale are pure noise that is not of physiologic interest. In theory, the dimension of pure noise should be the same as the dimension of the embedding space, in this case 10. However, for finite-length signals this is not generally the case; for signals that are sampled very rapidly, the dimension ν may decrease at small length scales in a purely artifactual way.

Rather than look for plateaus or peaks in the graph of $\nu(r)$ — as motivated by the search for chaos — it is sometimes more appropriate to choose length scales of interest and to average $\nu(r)$ between these length scales. In Figure 1, the length scales indicated by arrows correspond to the length that 0.5 percent of the pairs of points in the trajectory are closer than, and the length that 75 percent of the pairs of points are closer than.

This way of choosing the length scales of interest has several advantages. One advantage is that the dimension becomes independent of the variance of the signal (dimension is always independent of the mean level). This means that dimension and variance measure different aspects of variability. Another advantage is that the dimension reflects the dynamics at several length scales.[7] A third advantage is that by ignoring the most distantly separated points, the method becomes comparatively insensitive to glitches in the data such as that caused by ventricular ectopy. On the other hand, one loses the ability of the dimension estimate to identify deterministic chaos.

---

[7] A possible refinement of the method would be to use $\nu(r)$ at several values of r. This would allow the dynamics at different length scales to be monitored separately, but would also increase the amount of information that needs to be correlated clinically.

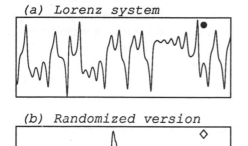

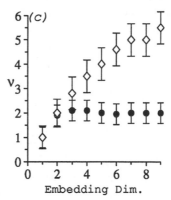

(a) *Lorenz system*

(b) *Randomized version*

(a) A signal derived from the x-coordinate of a mathematical chaotic system called the Lorenz equations.

(b) A random signal with the same power spectrum as the signal in (a).

(c) The dimension ν for the two signals in (a) and (b). The randomized signal has a higher dimension than the original.

Along a similar vein, Pincus et al.[8] have shown that calculating an *approximate entropy* with a small embedding dimension and fixed length scale allows a clinically useful characterization of heart rate variability even in cases where the theoretical entropy[9] does not appear to exist.

As mentioned above, the dimension and entropy can be calculated in a manner that makes them completely independent of the mean and variance of the heart rate and blood pressure signal. To what extent are the dimension and entropy independent of the shape of the power spectrum? Two signals with identical power spectra can have completely different

---

[8]S. Pincus, I. Gladstone, R. Ehrenkranz, "A chaos-related patternness measure for medical data analysis," xerox preprint
[9]i.e., the Kolmogorov-Sinai entropy

dimensions and entropies. Figure 2 shows a signal derived from a mathematical dynamical system called the Lorenz equations, along with a random signal with the same power spectrum. The random signal is more complex in that much more information is needed to specify that signal than the signal from the Lorenz equations. Calculation of the trajectory's dimension for different embedding dimensions highlights this difference in complexity. In addition, it is possible to combine nonlinear and linear techniques: for example, if one is interested in studying the effect of respiration on heart rate, one can extract a respiratory-band heart rate signal using linear filtering techniques. This respiratory-band signal can then be analyzed using nonlinear techniques.

Preliminary results of a study using dimension and entropy to quantify differences in heart rate and blood pressure variability between old and young subjects[10] indicate that these techniques can distinguish old from young in terms of cardiovascular regulation. This study supports a hypothesis that aging reduces physiological complexity. More important, the study indicates that dimension and entropy can be calculated in a robust manner on real heart rate and blood pressure data — that comparisons of dimension and entropy between individuals can be meaningful.

The eventual utility of nonlinear techniques such as dimension and entropy calculations for medical monitoring will depend on what clinical correlates are found for dimension and entropy. It is clear from work on linear analysis of heart rate variability that quantifying variability does have clinical utility — after more than a decade of research, variability monitors are just in the early stages of entering the market. At this early stage in the development of nonlinear techniques, it is hard to say what the final outcome will be.

As is often the case, there is a "Catch 22" in the introduction of novel monitoring technologies based on new principles. Physicians have had few means to measure variability and therefore have not been able to say what aspects of variability are of potential medical importance. Without knowing what to measure, it is difficult to develop effective ways of measuring it. We need simultaneously to develop methods for measuring variability and an understanding of the significance of what we're measuring. Nonlinear techniques may be useful in this endeavor.

---

[10]D.Kaplan, M.Furman, S.Pincus, S.Ryan, L.Lipsitz, A.Goldberger, *Aging and the complexity of heart rate and blood pressure dynamics*, in preparation.

TECHNIQUES FOR ASSESSING PULMONARY INPUT IMPEDANCE

B. Ha, C.L. Lucas, G.W. Henry, J.I. Ferreiro, R. Zalesak, B.R. Wilcox

University of North Carolina at Chapel Hill

CB #7065, 108 Burnett-Womack, Chapel Hill, NC 27599-7065

INTRODUCTION

Pulmonary input impedance (Z_{in}), which is estimated as the complex ratio of pulmonary artery pressure to pulmonary artery flow at each frequency, has proved to be useful in studying the physical characteristics of the pulmonary vascular bed. The relative position of the most salient feature of the pulmonary input impedance spectrum, namely, the first modulus minimum, has been evaluated in response to the surgical correction of congenital heart defects [1,2]. Current techniques for estimating the pulmonary input impedance spectrum are based on the fast Fourier transform (FFT) approach. While this approach is computationally efficient, the major limitation is that frequency resolution is limited by the heart rate of the data obtained from the patients and thus fails to resolve the frequency of the modulus minimum. In this study, an autoregressive moving average (ARMA) model of the pulmonary circulation, first proposed by Abutaleb [3], is developed as a nonparametric approach to the estimation of pulmonary input impedance. The goal of this work is to compare traditional techniques for estimating input impedance such as the cross-spectral averaging technique [2] and the random spectral technique [4] with the ARMA technique in terms of the quality of the Z_{in} estimates and, specifically, the frequency of the minimum point.

METHODS

Pulmonary artery pressure (P) and pulmonary artery flow (Q) data obtained from animal experiments were subjected to a time domain averaging procedure described previously [5] to yield noise-free P and Q harmonics as well as P and Q noise. Noise-free Q data were used as inputs in a simulation procedure involving a six-element lumped parameter model of the pulmonary circulation [5] to generate noise-free pressure output. P and Q noise data were added to their respective noise-free components to give data with

23

additive noise. Both types of data with additive noise were processed by three different estimation techniques--cross spectral averaging, random spectral technique, and the ARMA technique. The calculated spectra were compared to the actual modulus and phase spectra of the lumped model.

In the ARMA model, the relationship between presssure and flow at time $t = n\tau$ is assumed to be determined by the difference equation of the form:

$$P(k) = \sum_{i=1}^{n} a_i P(k-i) + \sum_{j=0}^{m} b_j Q(k-j) \qquad k = 0, ..., N$$

where $P(k-i)$ and $Q(k-j)$ are the measured pressure and flow, respectively. n and m correspond to the number of poles and zeros that exist in the system transfer function. a_i and b_j are the unknown parameters to be estimated that reflect the physical properties (such as resistance, compliance, inertance) of the system. To set up linear equations, the ARMA equation may be rewritten as:

$$z = F Q + v$$

where

$$z = [P_0 \quad P_1 \quad ... \quad P_N]^T$$
$$v = [Q_0 \quad Q_1 \quad ... \quad Q_N]^T$$
$$Q = [a_1 \quad a_2 \quad ... \quad a_n \quad b_0 \quad b_1 \quad ... \quad b_m]^T$$

and

$$F = \begin{vmatrix} P_{n-1} & P_{n-2} & \bullet & P_0 & Q_m & Q_{m-1} & Q_{m-2} & \bullet & Q_0 \\ P_n & P_{n-1} & \bullet & P_1 & Q_{m+1} & Q_m & Q_{m-1} & \bullet & Q_1 \\ P_{n+1} & P_n & \bullet & P_2 & Q_{m+2} & Q_{m+1} & Q_m & \bullet & Q_2 \\ P_{n+2} & P_{n+1} & \bullet & P_3 & \bullet & Q_{m+2} & Q_{m+1} & \bullet & Q_3 \\ \bullet & P_{n+2} & \bullet & \bullet & \bullet & \bullet & Q_{m+2} & \bullet & \bullet \\ \bullet & \bullet & \bullet & \bullet & \bullet & \bullet & \bullet & \bullet & \bullet \\ P_{N+n-1} & P_{N+n-2} & \bullet & P_N & Q_{N+m} & Q_{N+m-1} & Q_{N+m-2} & \bullet & Q_N \end{vmatrix}$$

Using a standard least squares procedure such as singular value decomposition [6], the coefficient vector Q is solved as:

$$Q = [F^T(N) \, F(N)]^{-1} \, F^T(N) \, z$$

The coefficients $a_1, a_2 \ldots, a_n$ and $b_0, b_1 \ldots, b_m$ determined by the method described above are substituted into the z-transform of the ARMA difference equation to calculate $Z_{in}(z)$ where $z = ejw = e^{2\pi\delta f\tau}$, $\delta f = 0.0833$ Hz, and $\tau = 0.02$ samples/sec.

RESULTS

Noise-free Studies

The cross-spectral technique as shown in Figure 1 determines impedance values at the 10 harmonics that are integral multiples of the heart rate. Thus, only the modulus spectrum for the data set with a heart rate closest to the fundamental driving frequency of the lumped model (1.80 Hz) exhibited a clear minimum point at the first harmonic. The cross-spectral averaging technique clearly lacked the frequency resolution necessary to delineate the minimum point for data with higher heart rates. The random spectral technique as shown in Figure 2 failed to resolve the frequency of the minimum point and generally yielded poor estimates on this noise-free data. The spectra obtained by processing data with an ARMA model of order (6,6) are shown in Figure 3. Results indicate that the minimum point of the modulus spectra and the overall spectra at the higher heart rates matched those of the true spectra, while the zero-crossing point of the phase spectra matched those of the true spectrum for all the heart rates given. For the data with the lowest heart rate, the estimates were underestimated for frequencies beyond the minimum point and overestimated below the minimum point.

Additive Noise Studies

In Figure 4, the minimum point of the modulus spectrum for data with a heart rate close to 108 beats/min clearly matched those of the true modulus spectrum. At higher frequencies, the real estimates obtained are equal to those of the true spectrum. The addition of noise to the data greatly improved the impedance values obtained by the random spectral technique (Figure 5). While the overall shape of the modulus spectrum was reproduced, the technique failed to resolve the frequency component of the minimum point. At 0-4 Hz, the phase estimates obtained were reasonable and the zero-crossing points of the phase curves corresponded to those of the true phase spectrum. Figure 6 indicates the tendency of the ARMA model of order (6,6) to underestimate the modulus values at frequencies higher than the frequency of the minimum point. At lower frequencies, the frequency at which the minimum point occurs was not resolved and modulus values were generally underestimated. The zero-crossing points of the phase spectra closely approximated that of true phase, and for the data set with the highest heart rate, the zero-crossing point matched that of the actual spectrum.

CONCLUSIONS

Under ideal conditions, results of this study indicate that the ARMA technique is capable of estimating continuous input impedance spectra and delineating the frequency of the modulus minimum for data at high heart rates. The zero-crossing points of the phase spectra were generally more consistent than the minimum point of the modulus spectra for the heart rates tested. For data with physiological noise, the random spectral technique yielded better estimates than the ARMA technique. However, neither technique resolved the frequency of the minimum point on the modulus plots. For both techniques, the zero-crossing points of the phase spectra were more consistent than the modulus spectra.

REFERENCES

1.   Lucas CL, Radke NF, Wilcox BR, et al. Maturation of pulmonary input impedance spectrum in infants and children with venrtrcular septal defect. Am J Cardiol 57:821-827, 1986.

2.   Neches WH, Park SC, Mathews RA, et al. Pulmonary artery impedance in evaluation of pulmonary vascular disease. Circulation 62(III):330, 1980.

3.   Abutaleb AS, Melbin J, Noordergraf A. Identification of time-varying ventricular parameters during the ejection phase. IEEE Trans Biomed Eng 33(3):370-378, 1986.

4.   Bendat JS and Piersol AG. Random Data. Wiley Interscience, New York, 1971.

5.   Ha B, Lucas CL, Henry GW, et al. Comparison of techniques for assessing pulmonary input impedance. Proc IEEE/EMBS 11:108-109, 1989.

6.   Press WH, Flannery BP, Teukolsky SA, et al. Modeling of data. In: Numerical Recipes: The Art of Scientific Computing. Cambridge University Press, Cambridge, 1989.

Supported by NIH R01-HL35389

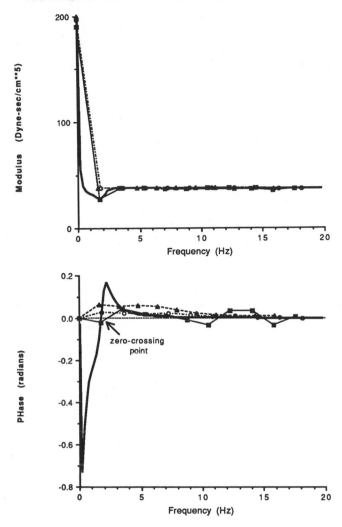

Figure 1. Input impedance spectra obtained by the cross spectral averaging technique. *Top*: modulus. *Bottom*: phase. Solid lines represent true spectra. Heart rates for the 3 data sets are given in parantheses. Open circle (107 beats/min), closed triangle (206 beats/min), closed square (248 beats/min).

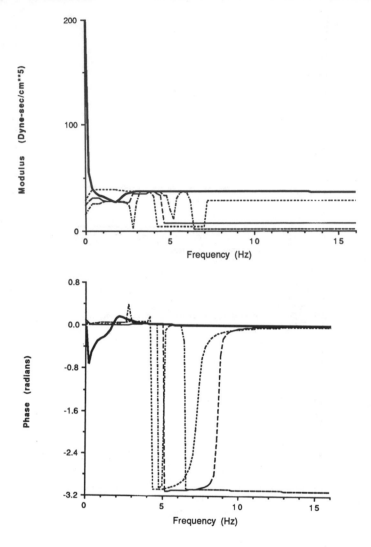

Figure 2. Input impedance spectra obtained by the random spectral technique. *Top*: modulus. *Bottom*: phase. Solid lines represent true spectra. Heart rates for the 3 data sets are given in parantheses. ———— (106 beats/min), —·—·· (190 beats/min), ----- (203 beats/min).

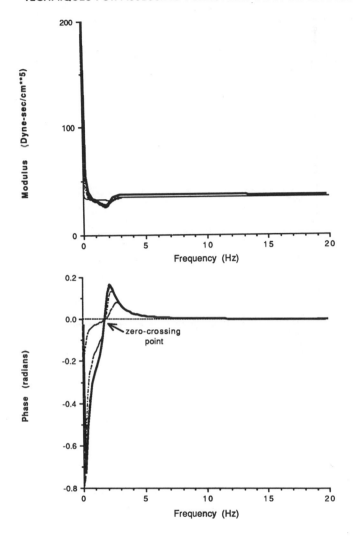

Figure 3. Input impedance spectra obtained by the ARMA technique. *Top*: modulus. *Bottom*: phase. Solid lines represent true spectra. Heart rates for the 3 data sets are given in parantheses. – – – – (105 beats/min), –·–·–· (192 beats/min), ----- (204 beats/min).

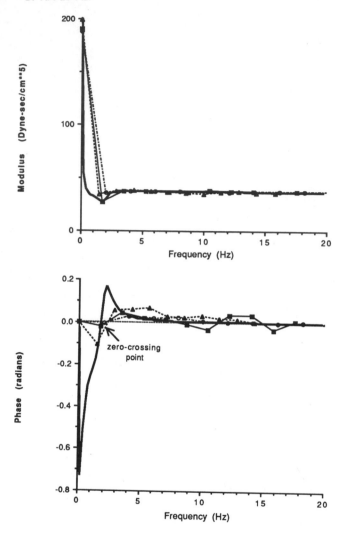

Figure 4. Input impedance spectra obtained by the cross spectral averaging technique. *Top*: modulus. *Bottom*: phase. Solid lines represent true spectra. Heart rates for the 3 data sets are given in parantheses. Open circle (107 beats/min), closed triangle (200 beats/min), closed square (240 beats/min).

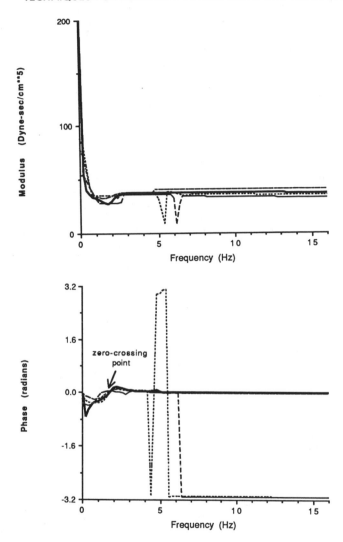

Figure 5. Input impedance spectra obtained by the random spectral technique. *Top*: modulus. *Bottom*: phase. Solid lines represent true spectra. Heart rates for the 3 data sets are given in parantheses. − − − − (106 beats/min), ⌐−−−· (196 beats/min), ----- (246 beats/min).

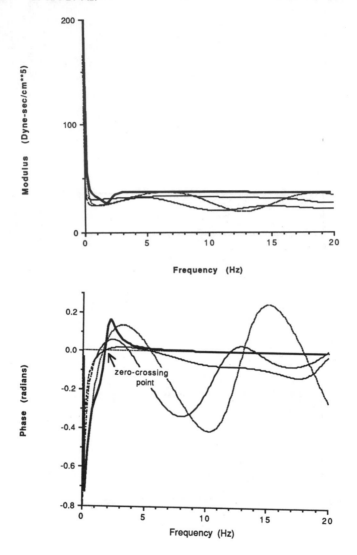

Figure 6. Input impedance spectra obtained by the ARMA technique. *Top*: modulus. *Bottom*: phase. Solid lines represent true spectra. Heart rates for the 3 data sets are given in parantheses. −−−− (108 beats/min), --·--·· (198 beats/min), ----- (245 beats/min).

THE MEASUREMENT AND CONTROL OF LUNG

MICROVASCULAR SURFACE AREA

Thomas R. Harris

Vanderbilt University

Nashville, Tennessee 37235

For some time we have been interested in the measurement of the amount of functioning microvascular surface area in the lung and the factors which control this function. Several questions can be posed about this system. What methods measure functioning capillary surface in intact animals or patients? How do vascular injuries affect the amount of capillary surface area? How can microvascular permeability and surface area be distinguished?

One approach to measuring surface area in the lung is the multiple indicator dilution method in which a mixture of materials is injected at the inflow to the lung and blood is sampled in the systemic circulation and analyzed (1). If the labeled materials are different in size and in their ability to cross the capillary barrier, a separation of concentrations will occur after the tracers have traversed the capillary bed. This separation is caused by a variation in the transport of the molecules across the capillary barrier and it can be used to measure a number of variables about the lung microcirculation.

The difference between an intravascular reference tracer and a diffusing tracer, termed the extraction, can be used to determine the permeability surface area (PS) for the escape of a small molecule such as ^{14}C-urea. A variety of mathematical models have been used to extract PS and other useful parameters from these data (1). In

33

addition, the mean transit time can be used to compute volumes--the total intravascular volume and the extravascular lung water volume.

How does lung injury affect surface area? If permeability is increased, PS is elevated (1). However, some injuries are more complex. Recently Bradley *et al.* (2) performed experiments in which glass beads were used to embolize isolated perfused dog lungs. Such embolization decreased urea PS as well as the flow conductivity of the lung vascular circuit (inverse of pulmonary vascular resistance). However, filtration coefficient in these experiments was not altered because the capillary beds continued to filter under the influence of elevated pulmonary venous pressure even though capillary surface was reduced. Indicator dilution PS was reduced by reduced area and a mismatch between perfused surface area and filtering surface area was also seen. This is an example of an injury in which permeability, surface area and filtration have a complex relationship.

Another complex injury is seen in the response of the sheep lung to endotoxemia (3). This particular model of injury causes an increase in lung lymph protein clearance, the product of lymph flow and protein lymph-to-plasma concentration ratio, under both Phase 1 (high pressure) reaction and Phase 2, increased permeability response. However, indicator dilution PS goes down in phase one and comes up to approximately baseline value in phase two. These results suggest that the lung somehow reduces perfused surface area after endotoxin infusion. Phase 2 PS may be a combination of reduced surface and increased permeability.

We have used the indicator dilution method to evaluate PS in adult respiratory distress syndrome (ARDS) patients (4). Pulmonary vascular resistance correlated inversely with urea PS in patients who did not reverse the x-ray and blood gas indicators of ARDS during 28 hours. This correlation did not exist in patients who reversed these signs. Their PS values were significantly lower. Apparently, the nonreversal patients have lost some ability to move blood away from injured areas and this may contribute to the severity of their disease.

The findings have suggested to us that it would be important to measure surface area and permeability separately. However, the indicator method always identifies the product of permeability and surface area. One approach to this separation is to use a transporting material which may be less sensitive to injury than purely hydrophilic tracers such as urea. One such material is butanediol which has both a lipophilic pathway through the capillary barrier and also a hydrophilic pathway. We hypothesized that urea may be more sensitive to true permeability change because of its confinement to the hydrophilic pathway. Whereas, the ability of butanediol to move directly through cell membranes may make it less sensitive to injury and more specific for surface area.

We studied this phenomena in isolated perfused dog lungs in which baseline indicator curves were injected and then the size of the lung was reduced (5). We also added alloxan to create increased capillary permeability. The results showed that the ^{14}C-urea PS is reduced when surface area is reduced but increased with alloxan. However, if PS for ^{14}C-butanediol is also determined in these experiments and we look at the ratio of the PS values (urea to butanediol), surface area cancels out. The ratio of the two PS values was significantly increased during the alloxan injury regardless of lung size. We concluded from this study that butanediol was less sensitive to alloxan injury than urea and could be used as a surface area marker. When urea PS was normalized to butanediol, surface area effects were eliminated. This method may be useful in analyzing the complex injuries discussed above.

In summary, many forms of lung injury reduce perfused surface area and increase permeability. Also, a method exists which separately accounts for permeability and surface changes. We are left with a question: is reduced surface area seen with these injuries a compensation for or as a manifestation of injury? If it is a compensation then pharmacological maneuvers should be used to reinforce this reduction. If it is simply part of the injury, the reduced surface area should be corrected pharmacologically. Further research is needed to clarify this situation.

ACKNOWLEDGEMENTS

This work was supported in part by the Public Health Service, N.I.H. Grant No. HL19153, SCOR in Pulmonary Vascular Diseases.

REFERENCES

1. Harris, T.R., Roselli, R.J. The exchange of small molecules in the normal and abnormal lung circulatory bed, pp. 737-791 IN: *Respiration Physiology: A Quantitative Approach*, Chang and Paiva, editors, Dekker, New York, 1989.

2. Bradley, J.D, Parker, R.E., Harris, T.R., Overholser, K.A. Effects of lobe ligation and bead embolization on Kf and permeability-surface area in the isolated dog lung. *J. Appl. Physiol.*, in review, 1990.

3. Bradley, J.D., Roselli, R.J., Parker, R.E., Harris, T.R. Effects of endotoxemia on the sheep lung microvascular membrane: A two-pore theory. *J. Appl. Physiol. 64*:2675-2683, 1988.

4. Harris, T.R., Bernard, G.R., Brigham, K.L., Higgins, S.B., Rinaldo, J.E., Borovetz, H.S., Sibbald, W.J., Kariman, K., Sprung, C.L. Lung microvascular transport properties measured by multiple indicator dilution methods in ARDS patients: A comparison between patients reversing respiratory failure and those failing to reverse. *Am. Rev. Resp. Dis. 141*:272-280, 1990.

5. Olson, L.E., Pou, A., Harris, T.R. Measurements of amphipathic and hydrophilic indicator-dilution tracers in the lung provide a surface area independent assessment of permeability changes. *FASEB J. 3*:A1140, 1989.

SUPERSONIC BONE CONDUCTION HEARING IN HUMANS:

A POTENTIAL NEW COMMUNICATIONS PATHWAY

R.A. Skellett, M.L. Lenhardt, and A.M. Clarke

Medical College of Virginia, Virginia Common-

wealth University, Richmond, VA 23298-0694

ABSTRACT

The upper limit of human hearing for air-conducted
signals is generally accepted to be 20,000 Hertz. There is,
however, experimental evidence from the last fifty years that
suggests man can hear two or perhaps three octaves higher,
but only by bone conduction. Bone conduction is that form of
hearing which results from direct vibratory stimulation of
the skull and ear. "Supersonic" bone conduction (ssBC)
hearing has been redefined in our laboratory as the sensation
of sound at frequencies above 20 kHz when a vibratory
stimulus is used. Recent studies have suggested that ssBC
may serve as a communication channel. To further explore
this possibility, speech signals were translated into the
supersonic range using the amplitude modulation-suppressed
carrier or double sideband (DSB) modulation technique. Once
modulated, the signals were then delivered through a
piezoelectric vibrator which eight normal subjects and two
deaf subjects held firmly to their mastoids. All of the
subjects were able to detect and perceive speech in the
supersonic range through this device. On average, the
percentage of correct responses was approximately 83% for the
normal subjects when using both a carrier frequency of 28.5
and 40 kHz. For the deaf subjects, the scores were lower,
20% and 30% correct for deaf subject #1 and #2, respectively.
An analysis of the type of perceptual errors committed
revealed a similar pattern of errors between the normal and
the deaf subjects. For both groups, the majority of errors
were consonant substitutions. These data confirm earlier

implications in the literature that ssBC hearing may support speech perception. Of greatest importance, this observation shows promise for an alternative aural rehabilitation approach to individuals with hearing impairment.

INTRODUCTION

Although ssBC perception has been studied for the past several decades, its use as a communication channel had not been previously investigated prior to the work done in our laboratory. For our initial study (Soltanpour, Skellett, et al., 1989), speech signals were translated into the supersonic range using amplitude modulation (AM) and delivered to the subjects through a piezoelectric transducer (vibrator). The carrier frequency used was 32 kHz. The word recognition test used was the NU Auditory Test #6 (List 1B), which is an open set test of 50 phonetically balanced words. Nine young normal adults (age 20-35) and three youths (age 12) were tested. The results yielded an average speech discrimination score of approximately 86% correct with a range from 80 to 96%. This suggests that young normal subjects can, in fact, detect and discriminate speech in the supersonic range.

THE NEW SYSTEM

Although discrimination scores were good using the AM system, subjects complained about the presence of the carrier contained within the signal being delivered. A new system was developed to alleviate the problem of carrier annoyance. In contrast to the previous work, the method used for translating the speech frequencies into the supersonic range was amplitude modulation-suppressed carrier or double sideband (DSB) modulation. The DSB circuitry was designed around the MC1496 Balanced Modulator/Demodulator integrated circuit. Again the test word set was delivered through a piezoelectric vibrator. The word discrimination test given was a modified version of the Word Intelligibility by Picture Identification exam (Ross and Lerman, 1971), which is a closed set test consisting of matrices with six picture cells. Twenty words were presented to the subjects using a 28.5 kHz carrier frequency and ten words using a 40 kHz carrier. Eight young normal adults (age 23-28) were tested. The test results using this system also demonstrate speech recognition in young normal subjects in the supersonic range. An average discrimination score of approximately 83% correct with a range from 70 to 90% was attained for both carrier frequencies employed. Perceptual errors were consistent with expected discrimination errors in that similar phonemes were usually confused.

When the normal subjects were asked to subjectively rate their confidence in their responses, the overall estimation of their performance was somewhat lower than the actual scores achieved. The short or nonexistent echoic memory associated with the signal is a potential source of this underestimation, which existed in all but two subjects. The average confidence percentage was 66.5% which shows that subjects had more confidence than the 50% level labelled as "sure". Also, regardless of the carrier frequency used, accuracy for discrimination was highly independent of their feeling of confidence in their response.

Once it was demonstrated that the ssBC/DSB device was capable of generating recognizable speech for normal subjects, the study was then supplemented to include an analysis of ssBC speech perception for two deaf subjects. The results of this preliminary experiment showed that it was possible for these two subjects to detect and discriminate the test words at a significant level. Deaf subject #1 achieved a score of 20% correct, and deaf subject #2 a score of 30% correct. Chance level for this test was 16.67%. An explanation for #2's slightly better performance could be related to his ability to detect ssBC signals more readily. #2's ssBC thresholds were approximately 15 dB of acceleration better than #1's. Thus, speech discrimination ability in the supersonic range seems to be directly related to the threshold of detection in that range.

While the performance of the deaf subjects was lower than the normals, an analysis of the speech errors suggested that they apparently used similar speech perceptual strategies. The error categories analyzed were consonant substitutions, vowel substitutions, consonant additions, and consonant deletions. For both groups, the majority of errors made were consonant substitutions. The differences that did exist between groups were likely due to the small number of words delivered rather than a different basis for perception. No notable discrepancy between the errors made by the two deaf subjects was revealed. Even though the deaf subjects differed in their degree of high frequency hearing ability (as determined from their pure tone air conduction audiograms), this did not seem to influence the type of acoustical confusions noted in the speech discrimination task.

ssBC AND THE ELDERLY

Another group of subjects that might benefit from the ssBC/DSB hearing device is the elderly. Although the speech discrimination test has not yet been demonstrated with this group, other data has been taken using a ssBC pure tone sinusoidal signal that would suggest that they should be able to detect and discriminate ssBC speech. A study was performed in order to compare the ssBC thresholds of a group of elderly and a group of normal subjects (Wang, et al., 1989). Twelve normal young subjects (age 23-35) and eight elderly subjects (age 55-84) were tested. Pure tone audiograms were given to the subjects in order to demonstrate that the elderly had   considerably higher air conduction thresholds than the normals in the audiometric range. On the other hand, when ssBC thresholds were obtained at 32 and 40 kHz, the results were similar. Apparently, presbycusis (hearing loss due to aging) is absent in the supersonic range. Thus, the elderly would be good candidates to use as subjects to prove the device effective.

CONCLUSION

The foundation for ssBC speech perception has been clearly established by the results of these initial investigations.

REFERENCES

M. Ross and J. Lerman, Word Intelligibility by Picture Identification. Pittsburgh: Stanwix House, Inc., 1971.

D.A. Soltanpour, R.A. Skellett, G. ter Horst, M.L. Lenhardt, and A.M. Clarke, "Speech perception above 20 kHz; A new mechanism for communication," in Abstracts of the Thirteenth Midwinter Research Meeting, (D.J. Lim, ed.), (St. Petersburg, Florida), Association for Research in Otolaryngology, February 4-8, 1990.

P.Y. Wang, G. ter Horst, R.A. Skellett, A.M. Clarke, and M.L. Lenhardt, "Supersonic bone conduction hearing in humans; Clues from the elderly," in Abstracts of the Twelfth Midwinter Research Meeting, (D.J. Lim, ed.), (St. Petersburg, Florida), Association for Research in Otolaryngology, February 5-9, 1989.

ETHICAL BEHAVIOR: A MECHANISTIC PERSPECTIVE

Stanley Rush, Ph.D.

The University of Vermont

Burlington, Vermont   05405

It seems a bit paradoxical that as society becomes more technically oriented, our interest in the humanistic concerns that are treated in the domain of ethics, increases. We see evidence of this in the professional journals, in the intense debates and confrontations over birth control, euthanasia, abortion, insider trading and drug use regulation. People's interest in ethics goes as far back in time as there have been philosophers. The written historical evidence is plentiful stretching back to the days of the Greeks' preeminence among civilized nations. Not just professional philosophers such as Socrates and Aristotle, but the authors of the Greek tragedies and even politicians dwelt on this puzzling aspect of philosophy.

Broadly speaking, there have been three principal themes running through the development of ethical concepts. One of these is the religious formulation in which our behavior is prescribed by what has been handed down by authority; revealed truths accepted on faith and interpreted by theologians. A second viewpoint is based on a secular approach which uses logical arguments as put forth by Descartes, Kant and St. Anselm for example or optimization criteria as advocated by Hume and Mill (Burtt,1939). More popular today, as people realize how difficult it is to make ethical decisions on the basis of a set of rules, is a personalized approach to making ethical decisions depending on the circumstances and on personal preferences. What is missing in all of these approaches is an "understanding" of how the ethical problem fits into nature's scheme. As engineers we are trained to understand problems on a scientific basis as a prerequisite to solving them. Correspondingly it is the objective of this paper to make a contribution to furthering our appreciation of the fundamental significance of ethics in the real world, as a prelude to considering ethical problems specifically.

Before starting this discussion, however, it might be desirable to consider just where engineers fit into the picture. It is clear from the examples already cited that new and more difficult ethical problems often arise due to the provision of new options by science and technology. It is reasonable to suppose that the people who created these options can contribute successfully to their proper selection. Moreover, under a certain set of very plausible assumptions, the concept of a "machine" is an excellent model for living organisms. Engineers who deal with machines, with their particular properties and their susceptibility to special techniques of analysis, may well have a favorable impact on the study of ethics and the resolution of its many difficulties. Another manner in which engineers might help stems from their "can do" professional attitude; engineers are trained to approach a problem, solve it and then move on to the next step. The liberal arts community, which has more or less had a monopoly on ethical analysis has, on the other hand, been disappointed so many times in their attempts to settle important issues, that the concept of closing out a problem may have become foreign to their thinking. (Some have in fact, rejected scientific approaches to philosophy as arriving too rapidly at their conclusions!) See Hampshire(1969). In addition to these specific arguments for engineers to be involved with ethical understanding, engineers have the same general interest in ethics as other members of the community.

The author's perspective on ethics focusses on the facts that ethical arguments have persisted over the ages and that we appear to be no closer to answers, or at least persuasive answers, now than we were 25,000 years ago and furthermore, that the problem of understanding the nature of what we call ethical problems is not a massively complex and remote one, as is, for example, that of explaining the birth and evolution of the Universe; ethical impulses are close to us every day and there ought to be a rationale for understanding them with which we can feel secure. Moreover, based on our history and experience, we can see that problems which have resisted resolution over a long period of time often are traceable to the use of hypotheses that have been accepted as true because

they seem so intuitively, but which eventually turn out
otherwise; one thinks of the effort the Greeks made to fit the
motions of the planets into a scheme in which all stars and
planets move in perfect circles.

To make a fresh start, it seems necessary to seek a
realistic set of relevant hypotheses which utilize reliable
and current scientific information. We have chosen to use
information from physics and evolution as guidelines for
"explaining" those mysteries of human psychology which
underlie ethics.

Physics is invoked in the most general of contexts. It
explains the very existence and properties of the earth in
which living matter arose from a sea of chemicals. Evolution
advises us with respect to how that initial life took shape
and evolved to greater diversity and greater complexity. Both
these theories suggest as a highly likely hypothesis that
human beings are machines, machines not designed by a
particular intelligence but by the trial , error, and
selection (based on survival in the broadest sense) scheme of
evolution. More and more, people are recognizing this as a
valid design approach, an approach incidentally which is very
often used as an aid to "intelligent" design as well. What
this argument leads to is the notion that ethical decisions
and ethical concepts have something to do with survival. This
concept is the basis for the very promising science of
sociobiology (Wilson, 1978).

The nature of the ethical problem appears to have two
principal characteristics. First of all, it involves
conflict. Conflict is readily defined in mechanical terms.
When a machine is motivated to perform two incompatible
actions simultaneously, there is by definition, a conflict.
In the typical automobile, the application of the brake and
the accelerator simultaneously represents such a conflict, one
which may have pathological consequences for the car. For
humans, choosing between a dinner out or an evening of tennis,
represents a conflict, but one easily resolved. Choosing
items from a dinner menu however, is for some people quite a
bit more difficult. But neither example represents an
"ethical" conflict. To understand the ethical conflict more
specifically, we need to add one more ingredient for which a
bit of evolutionary background is needed. It is helpful to
divide that history into two parts, the ages before our
ancestors came out of the sea, and after.

In the sea, life was not very demanding. Gravitational forces were cancelled by the buoyancy of the water, temperature fluctuations were small, and food and oxygen were all floating nearby. From the available evidence, it appears that animals then were not gregarious, they most often did not even have to care for their offspring. Their motivation was almost all "selfish" in the literal sense. (Even today, "shellfish are selfish.") After moving out of the water onto the land, when evolutionary pressures finally intensified, the much harsher land conditions apparently led to the evolution of social tendencies; taking care of the young, warning the group of impending danger, cooperative hunting and rearing, and so on. These social and altruistic tendencies are now inherited along with the self-preserving.The way things work out, simultaneous social and selfish motivations frequently appear in our daily lives. We sense the existence of this conflict because it is psychologically unpleasant. Sometimes these conflicts are life and death matters and the attendant psychological pain can be acute. A request from a beggar, rescuing at one's own peril a person who has fallen through the ice, filling out an income tax form, stopping to help a suspicious looking stranger apparently stranded by the side of the road, and so on are common examples.

In all such cases, the character of the problem is the same; a need to make a choice between personally advantageous versus socially responsible behavior. This is the second feature of an ethical problem.

Thus, there does exist a realistic and well-defined framework based on physics and evolution on which to build a model of the ethical situation, a model which conforms to the physical concept that we behave according to our physicochemical structure and that that structure has evolved according to evolutionary principles. We could on this basis easily formulate an ethical problem into a computer algorithm along the lines described by Rush(1988). Another interesting perspective on the same problem can be found in Asimov's (1950) novel "I Robot".

With this characterization of the essence of ethics in hand, we can begin to consider the issue of ethical activism, the development of behavioral guidelines for an entire society.

If we understand the nature of ethics according to the model described, can we then make "ethically correct" decisions on what ought to be done with respect to such subjects as euthanasia, abortion, welfare, providing food for starving infants in remote countries, and so on?

The "man is a machine" hypothesis suggests a negative answer. We manifestly cannot use the term "ought" in the context of a machine model, machines intrinsically, have no moral obligation to do anything. Neither do machines have any way of knowing what is "correct" in any absolute sense. They perform according to the way they are designed and built and that is the end of it.

The same must be said of humans, we perform according to the way we are constructed at the moment of acting and we are designed and constructed not only for personal survival, but more broadly for the perpetuation of our genes, some of which are shared with our relatives.

On specifically ethical issues, we also act according to the way we are motivated and that motivation reflects contributions from social and selfish tendencies selected over millennia and modified by the individual's personal experience or programming. An ethical choice between two clear alternatives,however, practically always involves a third possibility, namely procrastination. Thus, in an ethical crisis involving two active choices, there is almost always a "hesitation" option, that can serve to delay making a tough decision. During this interim, one can and often does spend the time rethinking the possibilities and collecting advice and data to determine the optimal action when the choice finally has to be made. It is the mental exercise in this deciding process that gives us the impression that we are acting freely, although the mechanistic model says otherwise, and contributes to making the action unpredictable. Thus, despite our use of forethought, our behavior results from the structures of our bodies at the moment of decision as is the case for a machine. From one point of view, the decision that is made is the outcome of an internal competition between our selfish and social instincts.

But this view applies specifically to the actions of one individual making a clear-cut choice between well-defined alternatives. Many ethical decisions on the other hand, involve social situations; a nuclear freeze, welfare abortion, capital punishment and the like. What one individual thinks about such matters scarcely counts. To accomplish a social objective, it is necessary to enlist many people who think, or can be persuaded to think, the same as the ethical activist. Bumper stickers are one manifestation of this activism. In large modern societies, there are many activist organizations that have mutually incompatible desires concerning the same issues, e.g., the pro-choice versus pro-life struggle. This line of thought reveals that the ethics of society are some rough average outcome of the preferences of a huge number of constituents. These outcomes are reflected in our laws and educational curricula. There are however, still no external moral standards imposed on us that tell us how to act in any given instance; we still behave as we are programmed to behave and to some extent we are programmed to obey society's laws generally. When people act as a majority of society wants them to, society calls that behavior ethical.

In summary, it comes down to this. The ethical problem is an artifact of our evolutionary history. We have a legal but no moral obligation to behave in a particular way that society dubs "ethical". On the other hand, each individual has as much right to see his or her views promoted in society as the next, social activism is thus "justified" in some sense. By these lights, we cannot blame the Hitlers, Genghis Khans, and Napoleons for the slaughters they instigated; those in society who find such actions sufficiently repulsive to act, must organize to prevent their occurrence as soon as the need is identified.

We close with several additional points. It is interesting that the views presented herein do not suggest any great changes in the way we conduct our lives; the analysis instead explains what is actually going on, ethically speaking, in the world. This fact is disappointing but we are at least provided with a meaningful and realistic interpretation of ethics. It explains, among other things, why philosophers have for so long looked in vain for rational ethical guidelines. If we ask why the simple view presented here, if true, has not been adopted long ago, the answer appears to lie in the peculiarities of human psychology; we tend to reject out of hand the consequences of the mechanistic hypothesis. But this same mechanistic hypothesis explains much more about philosophy than human ethics (Rush, 1989). It also elucidates many other philosophical issues; free will, responsibility for one's actions, the existence of God, the insanity defense, a rationale for sentencing criminals. In short, the basic assumptions regarding physics, evolution, and the mechanistic hypothesis can form an integrated world view of great explanatory power.

References

Asimov,I. (1950). I Robot. Doubleday, N.Y.
Burtt,E.A. (1939). The English Philosophers from Bacon to Mill. The Modern Library. N.Y.
Hampshire,S. (1969) in Symposium on J.L. Austin, K.T.Fann, ed. Routledge and Kagan Paul Ltd., London.
Rush,S. (1988). Catspaw: a model of animal behavior and learning. Behavioral Science, 33, 257.
Rush,S. (1989) Behavioral modelling: implications for philosophy. Proceedings of the Seventh International Conference on Mathematical and Computer Modelling. (In Press)
Wilson,E.O. (1978). On Human Nature. Harvard U. Press.

BIOMEDICAL ENGINEERING APPLICATIONS AND OPPORTUNITIES IN THE HUMAN LIFE SCIENCES PROGRAMS FOR SPACE STATION FREEDOM

Daniel William Barineau

KRUG Life Sciences Inc./Johnson Space Center

1290 Hercules, Suite 120, SS/P2
Houston, Texas 77058

The approaching construction of Space Station Freedom has presented the engineering community with a wide range of challenges to overcome. From a biomedical engineering (BME) standpoint, problems must be addressed concerning various equipment design parameters (function, weight, mass, volume, power, run time, commonality, amount of automation, data generation, etc.) as well as the condition of the equipment operators (astronauts). In terms of human life sciences, Space Station Freedom currently has three different programs to fill the requirements for; basic human research in microgravity, in-flight operational monitoring and countermeasures, and on-orbit acute clinical care. These are respectively the Space Biology Initiative (SBI), Biomedical Monitoring and Countermeasures program (BMAC), and the Crew Health Care System (CHeCS).

BIOMEDICAL ENGINEERING FIELDS

In defining the broad field of biomedical engineering, three convenient categories are: 1) the biomedical research scientist, 2) the biomedical design engineer, and 3) the clinical engineer in health care (Bronzino 1986). In the S.S. Freedom scenario, the three life sciences programs follow a similar pattern in function and organization, with SBI being the research program, BMAC being the operationally designed program, and CHeCS being the clinical program. Each of the three programs requires biomedical engineers to assist in the facility design and development process. Due to the different goals and objectives of the three programs, each contributes uniquely to the overall methodology for human life sciences on Station. Therefore, it is important that one understands the distinctions between the different programs in order to better identify the biomedical engineering applications.

51

SPACE BIOLOGY INITIATIVE

SBI is primarily a hardware development program that will provide lab equipment for a wide variety of scientific fields including human, animal, plant and exobiology research. The portion of the program that deals with human research is called the Space Physiology Facility. In an attempt to determine what types of scientific research will be done on Freedom, seven human life sciences disciplines and associated "reference experiments" were identified as those most likely for future study. The seven discipline level areas of study are; 1) Behavior and Performance, 2) Cell and Developmental Biology, 3) Cardiopulmonary, 4) Neuroscience, 5) Musculoskeletal, 6) Radiation, and 7) Regulatory Physiology. The hardware necessary to perform the "reference experiments" was identified, including items common with other Freedom programs. The resulting list consists of a suite of approximately 25 pieces of equipment which will be used primarily in support of human physiology research experiments. The initial Announcements of Opportunity (AO's) from NASA for performing SBI related experiments on Freedom are expected to be released within the next few years.

Biomedical research or bioengineering is the subset of biomedical engineering that is most applicable in the investigation of basic human biological processes. To this end, the human portion of SBI is designed to investigate the changes in basic mechanisms of human bodily functions in the microgravity environment. Some examples of the proposed investigations are; 1) individual and group factors that determine well-being and maintain performance during long-duration exposure to space, 2) gravity responsive mechanisms operant in regulating cellular and metabolic processes, 3) specific mechanisms underlying post-flight aerobic deconditioning and orthostatic hypotension, 4) endocrinological processes related to the onset and progression of muscle atrophy during microgravity, 5) model characterization of the changes in vestibular function during microgravity, 6) the biophysical basis of low dosage radiobiological effects of in-flight space radiation in relation to carcinogenesis, mutagenesis, and cataractogenesis, and 7) the pharmacologic consequences of pharmacokinetic changes in microgravity (Taylor 1988).

These are only a representative sample of the many types of investigations planned for mature S.S.F. operations. The period prior to the Assembly Complete phase (June, 1999) will be oriented less towards SBI type research and more towards operational programs. The BMAC and CHeCS programs both fall into this later category and will be in operation by the Permanently Manned Capability phase (July, 1997).

BIOMEDICAL MONITORING AND COUNTERMEASURES

BMAC is an operational program that is required to maintain optimum crew health and performance, to medically certify routine long-duration tours of duty, and to provide in-flight operational medical systems, procedures and countermeasures. The scientific background for BMAC is a set of maintenance goals that represent areas of astronaut health and performance in which problems are most likely to occur. These goals are divided into "operationally important problems" (19 total), each of which can adversely impact crew operations in-flight. Examples of these are; decreased ability to perform long duration tasks, visual dysfunction, changes in mood/motivation, altered pharmacologic activity, and increased occurrence of cardiac dysrhythmias. Equipment for monitoring and providing countermeasures to these problems was determined by generating a representative operational protocol and required equipment for each of the 19 problems. Because many of the resulting items already existed in other Freedom programs, the initial estimate of 72 pieces of BMAC provided hardware was reduced to 41.

As a program, BMAC is most like the description of a biomedical design engineer. Given a set of problems that are expected to occur during long-duration microgravity stays, the task BMAC has is to develop ways of monitoring physiologic and psychologic precursors to these problems, and formulate countermeasures to eliminate or reduce the detrimental effects. As previously mentioned, BMAC must be in place during the first period of continuously manned S.S.F. operation. This fact dictates a good working knowledge of the problems prior to any actual on-orbit operations. To solve this problem, an extensive survey was made of the available literature discussing human reactions to spaceflight. Direction in these areas came from a variety of sources, including past NACA missions (Mercury, Gemini, Apollo, Skylab, Spacelab, and Space Shuttle) and information from the soviet space program. In addition, information was found in a number of "strategic planning" reports for future NASA space biology efforts, and various conferences and meetings with scientists involved in the space-related human life sciences.

BMAC also has the responsibility to coordinate the access to human life sciences information that will result from S.S.F. investigations. This will be accomplished through an information tracking system that will allow the history and status on any given datum to be readily accessible. The BMAC program will therefore have a working knowledge of the operations of all of the human life sciences programs, and can minimize the duplication of S.S.F. efforts.

CREW HEALTH CARE SYSTEM

CHeCS is actually made up of three different Space
Station facilities; the Environmental Health System (EHS),
the Exercise Countermeasures Facility (ECF), and the Health
Maintenance Facility (HMF). The EHS is responsible for
conducting environmental health assessments on Station.
These include sampling, sample processing, and analysis of
air, water, internal surfaces, internal vibroacoustics, and
internal and external radiation environments. The ECF
provides equipment and techniques for in-flight exercise and
physiological monitoring. The goal is to prevent physical
deconditioning from long-term exposure to the microgravity
environment. The HMF provides diagnostic, and therapeutic
medical support for injured or ill crewmembers. Some of the
capabilities of this facility include a pharmacy, in-flight
surgery, fluid and hyperbaric therapy, and medical life
support. In total, the CHeCS program has approximately 100
hardware items within the three component facilities.

Of the three BME subsets, clinical engineering is
closest in nature to the CHeCS program. The standard
clinical care needs provided by ground-based hospitals are
very similar to those of CHeCS. The primary differences are
that astronauts are above average in health and fitness, and
the clinical care technologies must be applied in a
microgravity environment with confined mass, power, volume,
and operational envelopes. Therefore, clinical engineers
must provide the capabilities that the various medical
situations require, while designing their systems within the
given set of engineering constraints.

These design requirements have generated the need for
CHeCS concept testing prior to operation on the Station. A
large number of medical operations and investigations are
being examined on the Shuttle and the zero-g KC-135 airplane.
The key technical challenges that have been identified are;
1) the development of non-invasive diagnostic techniques for
S.S.F. usage, 2) conversion of gravity dependent equipment to
enable it to function in microgravity, 3) x-ray quality
imaging for use in ground-based health assessment, 4) the
performance of exercise countermeasures without serious
perturbation of the S.S.F. microgravity environment, and
5) active radiation monitors for radiological health risk
assessment (White 1989). The answers to these questions will
greatly affect the operations of the other two life sciences
programs. Therefore, an important part of CHeCS, BMAC, and
SBI are the interactions and cooperation between the facility
and program developers of each program. The final goal of
this effort is a well-coordinated, goal-oriented, non-
overlapping set of life sciences programs.

BIOMEDICAL ENGINEERING APPLICATIONS

Biomedical engineers are required in all levels of operation within the three aforementioned programs. Early on in the program development, applications for biomedical engineers include; providing an interface between the program scientists and the hardware engineers, ensuring that information about applicable new technologies is available to the researchers and scientists, determining resource impacts for the planned investigations, prototype design and development, and aiding in the creation of ground based studies for technology verification. During in-flight operations, analysis of experimental data and experiment procedure support would become primary BME functions, along with experiment evolution as the human-in-microgravity knowledge base increases. Post-flight data collection and analysis would also be an area where BME applications would be found.

All of the tasks performed by biomedical engineers must be done in conjunction with either NASA personnel and/or private contractors. Since NASA has divided the S.S.F. development, construction, and integration between the various centers scattered around the country (JSC, KSC, MSFC, AMES, etc.), involved biomedical engineers must first identify who does what for the various Station programs. Subsequently, they need to know how to communicate their ideas to these people in order to effectively represent their projects. This necessitates a basic understanding of many of the systems that will be directly or indirectly affected by the life sciences operations. Examples of these are; 1)gas, fluid, and waste management systems, 2)power, computer, and information systems, 3)S.S.F. provided support equipment for lab use, 4)crew availability and capabilities, 5)logistics and resupply parameters, 6)station-station and station-ground communication lines, and 7)the environmental monitoring systems.

Because of these and other expected interactions between space-based technology and human beings on Freedom, the opportunities and applications for biomedical engineers are numerous. One of the most valuable of these is the ability to communicate between the two different spheres of knowledge (engineering and biological science) during the creation and operation of S.S.F. The final judgement on the influence of these engineers, will be the Space Station's usefulness as a productive life sciences research environment for the for the next 30 years.

References

1. Bronzino, J.D. 1986. *Biomedical Engineering and Instrumentation: Basic Concepts and Applications.* Boston: PWS.

2. Taylor, G.R., ed. 1988. *Space Station Freedom Human-Oriented Life Sciences Research Baseline Reference Experiment Scenario.* NASA Doc. JSC-32084.

3. White, T.T. 1989. *Presentation to Headquarters on the Space Station Crew Health Care System.* Unpublished.

II.

MATHEMATICAL MODELS
IN MOLECULAR BIOLOGY AND PHYSIOLOGY

HOW TO SIMULATE THE QUANTUM-MECHANICAL SUPPRESSION OF HIGH-FREQUENCY MODES IN CLASSICAL MOLECULAR DYNAMICS COMPUTATIONS

Charles S. Peskin and Tamar Schlick

Courant Institute of Mathematical Sciences

251 Mercer Street, New York, N.Y. 10012

The purpose of this talk is to describe a molecular dynamics method which combines two old ideas in a new way. These ideas are the backward-Euler method for the solution of stiff ordinary differential equations and the Langevin-equation approach to the maintenance of thermal equilibrium. The new feature is that the Langevin collision frequency and the backward-Euler dissipation rate are balanced in such a way as to produce a definite cutoff frequency which can be set equal to kT/h to simulate the quantum-mechanical suppression of high-frequency modes. The resulting BACKWARD-EULER/LANGEVIN method has been tested on two physical problems for which both the quantum mechanical and classical results are known: a system of coupled harmonic oscillators and a diatomic molecule. In both cases, the method produces results that correspond much more closely to those of quantum than to those of classical statistical mechanics, despite the classical character of the method. Implementation of the method in the case of large biological molecules will also be discussed.

ACKNOWLEDGEMENT: This work is supported by the National Science Foundation under research grant ASC-8705589.

REFERENCES:

1. Peskin, C. S. and Schlick, T.:   Molecular dynamics by the backward-Euler method.   Commun. Pure & Appl. Math. 42: 1001-1039, 1989

2. Schlick, T. and Peskin, C. S.: Can classical equations simulate quantum-mechanical behavior: a molecular dynamics investigation of a diatomic molecule with a Morse potential. Commun. Pure & Appl. Math. 42: 1141-1163, 1989

3. Peskin, C. S.: Analysis of the backward-Euler/Langevin method for molecular dynamics. Commun. Pure & Appl. Math. (in press)

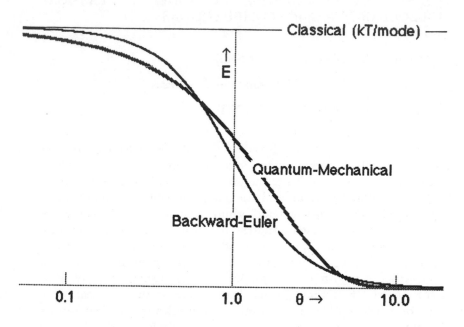

Fig. 1. Partition of mean thermal energy among the vibrational modes of a system of coupled harmonic oscillators ($\theta = h\nu/kT$). Note qualitative agreement between the backward-Euler and the quantum-mechanical results, both of which are in sharp disagreement with the classical equipartition theorem (E independent of θ).

INFORMATION PROCESSING IN THE AUDITORY BRAINSTEM

Michael C. Reed

Jacob J. Blum

Department of Mathematics
Duke University
Durham, NC   27706

Division of Physiology
Department of Cell Biology
Duke University Medical Center
Duke University
Durham, NC   27710

The purpose of this talk is to discuss general questions about information processing in the auditory brainstem in the context of a specific model for the encoding of azimuthal position in the lateral superior olive.

A great deal of information has been obtained from single unit recordings in the auditory system in response to specific external stimuli [1], [2]. Such recordings are not easy to interpret since one is seeing not just the properties of the neuron where the recording is taking place but also the properties of all the neurons between the cochlea and the recording site. In our model for the lateral superior olive, neurons from the antero ventral cochlear nucleus (AVCN) project to columns of cells in the lateral superior olive (LSO); see Figure 1. The 40 AVCN cells are assumed to respond directly, with simple response curves, to sound stimulation on the ipsilateral side, but the response is graded because the thresholds of the AVCN cells increase as i goes from 1 to 40. As indicated in Figure 1, the AVCN neurons with lower thresholds tend to make connections with the lower part of the LSO column and the ones with higher thresholds with the higher part of the column. These connections are excitatory and because of the variation of threshold the amount of incoming excitation to the cells in the LSO column decreases as one proceeds up the LSO column. If one were to record from the cells in the LSO column one would detect an increase in LSO cell threshold (to external stimuli) as one moved up the column even if the LSO cells themselves are identical. The graded responses of the LSO cells in this model are not intrinsic properties, but system properties reflecting (in

59

this case) the difference in thresholds of the AVCN cells and the serial nature of the projections from the AVCN to the LSO.

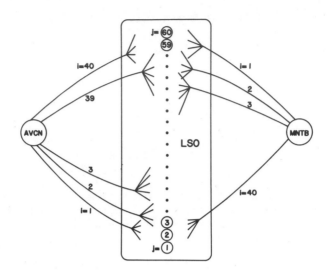

Figure 1.
Connections between the AVCN, MNTB, and LSO.

The point is that one cannot really interpret the experimental data of single unit recordings without an understanding of the properties of the systems leading to the neuron.

How is one to understand these system properties? By reasoning backwards from the single unit data? By using detailed anatomical studies? By utilizing guesswork based on the perceived "function" of the system? Surely the answer is that one must use all three methods together. The anatomy is too complicated to be investigated without some apriori ideas based on function. Top down models based solely on function are not likely to give understanding of real physiological systems. All three methods must be used to construct theoretical models of how the system works. These models must be quantitative, because we know that system behavior depends not only on how the system is connected up and the strengths of the connections, but also because one must be able to derive quantitative conclusions from the models so that they can be tested by machine computation and compared to experiment.

The second point is that theory based on function can provide a basis on which to make structural hypotheses about local anatomy and physiology, hypotheses which can be tested experimentally. The ipsilateral LSO receives inhibitory input from the ipsilateral medial nucleus of the trapezoid body (MNTB) which is stimulated by the sound at the contralateral ear; see Figure 1. For this reason it has been thought for some time that the LSO computes (or encodes) azimuthal position on the horizon on the basis of interaural intensity difference. In our model, we assume that the MNTB neurons have a range of thresholds and that they project serially onto the LSO column in the opposite way from the AVCN projections, that is neurons with low thresholds project toward the top of the column and neurons with high thresholds toward the bottom. For given intensities at the two ears one obtains monotone decreasing incoming excitation and monotone increasing incoming inhibition as one goes up the LSO column. If one makes the simple assumption that an LSO cell fires if and only if the incoming excitation is greater than the incoming inhibition, then all the cells in the column below the crossover point of the excitation and inhibition curves will be firing. Position on the horizon is encoded essentially by the crossover point which moves up and down depending on the interaural level difference which in turn depends on azimuthal location of the sound source. We came to this model structure by following anatomical hints but also by the following theoretical considerations. However excitation and inhibition interact at the LSO they must do so (if the LSO is to code for azimuth) in a way that is independent of absolute sound level; only the interaural difference should matter. As the absolute intensity goes up both the excitation and inhibition curves will rise but the crossover point might stay the same. In fact, our experience with the model has shown that firing output on the LSO column is quite independent of absolute intensity, in agreement with the results of psychophysical experiments.

We have investigated the model extensively by machine computation and have found that it encodes azimuthal location in a regular way and is quite stable under variations of parameters [3]. It is also consistent with known anatomy and physiology. This does not prove that the structural scheme we have proposed is right. But the model is quantitative and explicit and as such makes quantitative predictions which can be checked.

We have not mentioned the details of the connectional

schemes used in the model. We tried three types of schemes.
In the first we rigidly specified all the connections assuring
even distributions of synapses. In the second we allowed the
incoming AVCN and MNTB neurons to make stochastic choices about
where to form synapses. In the third, or trophic scheme, we
allowed the target LSO cells to make stochastic choices about
which parent AVCN and MNTB cells were allowed to synapse on
them. The first, hard-wired scheme worked well but is un-
physiological. The second worked quite poorly, but the third
worked excellently lending some support to the view that
trophic interactions play an important role in neural information processing [4].

References

[1]   Irvine, D.R.F. (1986). "The Auditory Brainstem," in H.
      Autram, D. Ottoson, E.R. Perl, R.F. Schmidt, H. Shimazu,
      and W.D. Willis (eds.): Progress in Sensory Physiology 7,
      New York, Springer-Verlag, pp. 1-21.

[2]   Pollak, G.D. and Casseday, J.H. (1989). The Neural Basis
      of Echolocation in Bats. Springer-Verlag, Berlin, 143
      pages.

[3]   Reed, M.C. and Blum, Jacob J. (1990). "A Model for the
      Computation and Encoding of Azimuthal Information by the
      Lateral Superior Olive," to appear in J. Amer. Acoust.
      Soc.

[4]   Purves, D., (1988). "Body and Brain: A Trophic Theory
      of Neural Connections," Cambridge, Harvard University
      Press, 231 pages.

A KINETIC HAIRPIN TRANSFER MODEL FOR DNA REPLICATION IN PARVOVIRUSES

Katherine C. Chen and John J. Tyson

Department of Biology
Virginia Polytechnic Institute and State University
Blacksburg, Virginia 24061

All linear DNA molecules face special problems in replicating their 5′ ends, as DNA polymerases add nucleotides
only to pre-existing strands with free 3′-OH groups. Parvoviruses, a group of small animal viruses with a linear single-stranded DNA genome, cope with this problem by having
palindromic terminal sequences that fold back on themselves
to form hairpin structures essential in priming DNA replication. After replication the hairpin structure at the co-
valently closed end of the double stranded DNA molecule must
be opened. This is accomplished by a site-specific nicking
enzyme that cuts the parental strand at the 5′ end of the
palindromic terminal sequence, transferring the palindrome
to the daughter strand and leaving a free 3′-OH group on the
parental strand. By copying from the free 3′-OH a complementary copy of the terminal palindrome is added to the parental
strand. If the palindrome is imperfect, the new terminal
sequence on the parental strand is slightly different from
the original sequence. The two reverse-complementary terminal sequences are referred to as "flip" and "flop". The 3′
end of the daughter strand, which is complementary to the 5′
end of the parental strand, faces exactly the same problem
when it is replicated. Therefore, there are "flip" and
"flop" sequences at both ends of all linear parvoviral DNA
molecules. This mechanism for replicating linear DNA molecules is called hairpin transfer.[1]

The relative abundances of the flip and flop sequences
at each end of the DNA molecules involved in parvoviral replication can be measured. For instance, for bovine parvovirus
(BPV) which encapsidates 90% minus strand and 10% plus strand
DNA, the relevant flip/flop ratios are[2]:

```
left  (3') end of minus strand, 10:1
right (5') end of minus strand,  1:1
left  (5') end of  plus strand,  1:1
right (3') end of  plus strand,  1:2
left  end of double-stranded replicating forms, 10:1
right end of double-stranded replicating forms,  1:2
```

These ratio are, in our theory, a reflection of the relative rates of hairpin transfer at the different ends of these DNA molecules.[3,4]

To specify the rates of DNA replication we introduce the following terminology: PV, PC, R for minus strands, plus strands, and replicating forms, respectively, and subscripts i and o for for flip and flop conformations at the termini. There are twelve different DNA species to track in the simplest model. Late in an infection (Fig. 1), when nearly all single-stranded DNA molecules are being encapsidated into virion particles (and thus unavailable for replication), the differential equations describing the system are

$$dR_{ii}/dt = -(k_{li} + k_{ri})R_{ii} + k_{ro}R_{io} + k_{lo}R_{oi}$$

$$dR_{io}/dt = k_{ri}R_{ii} - (k_{li} + k_{ro})R_{io} + k_{lo}R_{oo}$$

$$dR_{oi}/dt = k_{li}R_{ii} - (k_{lo} + k_{ri})R_{oi} + k_{ro}R_{oo}$$

$$dR_{oo}/dt = k_{li}R_{io} + k_{ri}R_{oi} - (k_{lo} + k_{ro})R_{oo}$$

$$dPV_{ii}/dt = k_{ri}R_{ii}, \qquad\qquad dPC_{ii}/dt = k_{li}R_{ii}$$

$$dPV_{io}/dt = k_{ro}R_{io}, \qquad\qquad dPC_{io}/dt = k_{li}R_{io}$$

$$dPV_{oi}/dt = k_{ri}R_{oi}, \qquad\qquad dPC_{oi}/dt = k_{lo}R_{oi}$$

$$dPV_{oo}/dt = k_{ro}R_{oo}, \qquad\qquad dPC_{oo}/dt = k_{lo}R_{oo}$$

The four rate constant, $k_{li}, k_{lo}, k_{ri}, k_{ro}$, describe the rates of hairpin transfer at the four different 3' termini: the left ends (flip or flop) of the minus strands, and the right ends (flip or flop) of the plus strands.

This system of ordinary differential equations quickly reaches a steady state in which the relative DNA abundances become constant, and the characteristic flip:flop ratios become

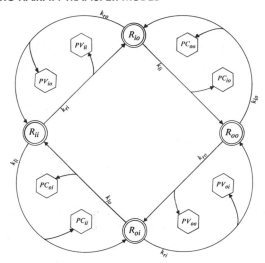

Fig. 1 Network of parvoviral DNA replication at packaging stage.

left   end of minus strand, $k_{lo} : k_{li}$
right end of minus strand, 1:1
left   end of plus strand, 1:1
right end of plus strand, $k_{ro} : k_{ri}$
left   end of replicating form, $k_{lo} : k_{li}$
right end of replicating form, $k_{ro} : k_{ri}$
minus/plus strands encapsidated, $k_{ri}k_{ro}(k_{li} + k_{lo}) : k_{li}k_{lo}(k_{ri} + k_{ro})$

Choosing $k_{lo} : k_{li} = 10:1$, $k_{ro} : k_{ri} = 1:2$, $k_{ro} : k_{lo} = 1:1$, we obtain the DNA distribution observed for BPV. Notice that the two flip:flop ratios constrained by theory to be 1:1 are also observed to be 1:1. Also, as theory predicts, the flip:flop ratios at the left and right ends of replicating forms are identical to the ratios at the left end of minus strands and the right end of the plus strands, respectively. These constraints are satisfied not only by BPV, but also by three other parvoviruses for which the DNA distributions are known.[3]

The key to understanding parvoviral replication, in our opinion, is the concept of kinetic constraints on the rates of individual steps in the replication pathway.

REFERENCES

1. T. Cavalier-Smith, *Nature (Lond)* 240:467-470 (1974).

2. K. C. Chen, B. C. Shull, M. Lederman, E. R. Stout and R. C. Bates, *J. Virol.* 62:3807-3813 (1988).

3. K. C. Chen, J. J. Tyson, M. Lederman, E. R. Stout and R. C. Bates, *J. Mol. Biol.* 208:283-296 (1989).

4. J. J. Tyson, K. C. Chen, M. Lederman and R. C. Bates, *J. Theor. Biol.* (in press).

ACKNOWLEDGMENTS

This work was supported by grant GM36809 from National Institute of Health and grant MV-220 from American Cancer Society.

A COMPARISON BETWEEN THE FORMAL DESCRIPTION OF REACTION AND NEURAL NETWORKS: A NETWORK THERMODYNAMIC APPROACH

Donald C. Mikulecky

Departments of Physiology and Biology

Virginia Commonwealth University

Richmond, VA 23298-0551

BITNET:MIKULECKY@VCUVAX

It does not take any effort to recognize that biochemical pathways are among the many kinds of networks of which living systems are made [1-25]. These reaction networks are so clearly different from the neural networks which now command considerable attention, that a serious comparison seems almost frivolous. However, in his work on complex systems, Rosen [26-28] points out some similarities which deserve further attention.

To examine these similarities in a systematic manner, network thermodynamics will serve as a tool for the formal description of both. Network thermodynamics is a field based on the dual nature of living systems, namely that as **systems** they are as much a function of their topology as the more readily recognized parameters: rate constants, volumes, pressures, electrical potentials, etc. [8-21,23-25] These latter properties are captured by the network elements used to "connect together" a representation of the system. The former is more subtle, since it is the pattern of those connections itself. It is, of course, the topology which is most frequently sacrificed by modern reductionist methods. One goal of network thermodynamics is to salvage the information gained using reductionist methods and synthesize some holistic composite of the parts. With a great deal of luck, this may begin to resemble the intact living system we started with.

In this spirit, it is possible to synthesize elaborate biochemical networks using individual reactions as the building blocks. With a number of techniques it is also possible to perform robust simulations of these biochemical networks. In Peusner's network thermodynamics [18,19], it is customary to represent each chemical species as a network node and the reaction pathways between species as branches containing elements which specify the reaction kinetics as a reaction flow. The result of such an endeavor might be represented by a linear graphical diagram which can be as complex as a small integrated circuit [29]. This diagram is complicated enough to arouse images of computer chips or neural networks. Its caption in a well known text of cell biology is "some of the reactions in a living cell". At this level of representation, the biochemistry is encoded as a linear graph. If we wish more information we use network elements in each branch to encode the type or reaction, and augment each node with either a source or volume capacitor [14-17,20,21,24,25,31] to mimic the experimental or observed conditions. Sources clamp or control the concentrations of the various chemicals, while the volume capacitors allow the concentrations to float and achieve the same conservation statement in a discrete representation as the continuity equation does in a continuum representation. The choice of these circuit elements is motivated by two considerations: First the practice of using the circuit simulator, SPICE, as an all purpose simulator, and second the work of Mason [32] which introduced the unistor as a model for a particular kind of circuit element which was to be the building block for non reciprocal networks. In actuality, in these biochemical networks, the unistor is merely the general element to represent

first order kinetic steps, either in reaction
schemes or compartmental systems [8-13,17,21,24,
25,31,32]. More complicated kinetics such as
Michaelis-Menten or higher order reactions are
handled by using controlled sources with polynomial
formats for their constitutive relations. An
example is the model of folate metabolism used for
the study of chemotherapeutic compounds in cancer
treatment [17,23,24].

In the simplest interpretation of neural
networks, the nodes are analogous to the cell
bodies of the neurons and the branches represent
the axons and dendrites making the connections
between the cells [33-36]. All the axons do is to
transmit an all or none signal to the synapse and
there, in fact, the signal becomes biochemical once
more. Although the synaptic event is rich in its
detail, artificial neural networks tend to adopt a
limited number of stereotypical transfer functions
to model these events. What is preserved is a kind
of temporal summation which becomes the central
part of the network's function. It is no wonder,
then, that at first glance the two systems might
seem dissimilar. This is even a more likely
impression if artificial neural networks' topology
is considered, since often all possible connections
between neurons are modeled. The only way that
topology becomes a factor is through the modulation
of the connection strengths between individual
neurons. Thus, in these artificial neural
networks, the constitutive relations are variable
and in fact alter the functional topology to a
great extent. If the artificial neural network is
at all representative of the real thing, then the
neural network has a constantly changing set of
constitutive relations. We might also
appropriately model these as controlled sources as
we did the constitutive laws for biochemical
networks.

Once we examine the properties of a
biochemical network in some more detail, it is
clear that the constitutive relations are also
variable! These variations are in the form of
inhibition and/or activation patterns. In fact the

example of folate metabolism given above, is rich
in such possibilities [17,23,24].

Rosen [26-28] focuses on the patterns of
inhibition/activation and further modifications of
these patterns by agonists/antagonists to establish
a very important mathematical feature that both
types of system have in common. This feature is
the demonstration that in order for these systems
to be characterized by the usual notion of states,
these patterns of modulation must be reciprocal in
their manifestation, a condition which is probably
never met. This idea alone is enough to make Rosen
sure that the classical Newtonian paradigm does not
rigorously hold for these systems. This same idea
was recognized by Mason for the electronic systems
he modeled with unistors. Network thermodynamic
models of both biochemical and neural networks are
consistent with this important attribute.

A final thought is that biochemical events are
becoming more widely recognized as information
transmitting processes in their own right outside
their role in synaptic transmission within the
nervous system. Modern molecular biology has
identified "second messengers" and "signal
transduction" mechanisms that extend the
traditional role of hormones as information
carriers to a more local level [37].

REFERENCES

1. Cornish-Bowden, A. (1979) <u>Fundamentals of
    Enzyme Kinetics</u>, Butterworths, London.

2. Garfinkel, D. (1984) Modeling of Inherently
    Complex Biological Systems: Problems,
    Strategies, and Methods, <u>Math.Biosci. 72</u>:131-
    140.

3. Heckmer, H. C. and B. Hess (1972) <u>Analysis and
    Simulation of Biochemical Systems</u>, Elsevier,
    Amsterdam.

4. Heinrich, R., M. Rapoport, and T. A. Rapoport
    (1977) Metabolic Regulation and Mathematical
    Models, Prog. Biophys.& Mol. Biol. 32:1-82.

5. Hill, T.L. (1977) Free energy transduction in
    biology, Academic Press, N.Y.

6. Hill,T.L. (1982) Linear Onsager coefficients
    for biochemical kinetic diagrams as one-way
    cycle fluxes, Nature 299:84-86.

7. King, E. L. and C. Altman, (1956) A schematic
    method of deriving the rate laws for enzyme
    catalyzed reactions, J. Phys. Chem.
    60:1375-1378.

8. Mikulecky, D. C. and S. R. Thomas (1979) Some
    network thermodynamic models of coupled,
    dynamic, physiological systems, J. Franklin
    Inst. 308: 309-325.

9. Mikulecky, D.C.(1983) A Network Thermodynamic
    Approach to the Hill-King & Altman Approach to
    Kinetics: Computer Simulation. In: Membrane
    Biophysics II: Physical Methods in the Study
    of Epithelia (M. Dinno A.B. Calahan and T.C.
    Rozzell eds.) A.R. Liss, N.Y. pp 257-282.

10. Mikulecky, D.C.(1984) Network Thermodynamics: A
    Simulation and Modeling Method Based on the
    Extension of Thermodynamic Thinking into the
    Realm of Highly Organized Systems, Math.
    Biosci. 72:157-179.

11. Mikulecky, D.C. and F. A. Sauer (1988) The role
    of the reference state in nonlinear kinetic
    models: Network thermodynamics leads to a
    linear and reciprocal coordinate system far
    from equilibrium." J. Math. Chem. 2:171-196.

12. Mikulecky, D.C. (1987) Simulation of the
    Mitochondrial Energy Transduction Mechanism: Is
    Chemiosmosis Local or Global? in Simulators IV,
    94-96.

13. Mikulecky, D. C. (1987) Topological Contributions to the Chemistry of Living Systems. in <u>Graph Theory and Topology in Chemistry</u>, (R. B. King and D. H. Rouvray, eds.) Elsevier, N. Y. pp. 115-123.

14. Oster, G.F., A. Perelson and A. Katchalsky (1973) Network thermodynamics:dynamic modelling of biophysical systems, <u>Quart. Rev. Biophys.</u> <u>6</u>:1-134.

15. Peusner, L. (1970) <u>The principles of network thermodynamics and biophysical applications</u>, Ph. D. thesis, Harvard. Univ., Cambridge, MA. [Reprinted by Entropy Limited, South Great Road, Lincoln, MA 01773,1987]

16. Schnakenburg, J. (1977) <u>Thermodynamic network analysis of biological systems</u>, Springer-Verlag, N.Y.

17. Seither, R. L., D. F. Trent, D. C. Mikulecky, T. J. Rape and I. D. Goldman (1989) Folate Pool Interconversions and Inhibition of Biosynthetic Processes after Exposure of L1210 Leukemia Cells to Antifolates, <u>J. Biol. Chem.</u> <u>264</u>: 17016-17023.

18. Peusner, L. (1982) Global Reaction-diffusion Coupling and Reciprocity in Linear Asymmetric Networks, <u>J. Chem. Phys. 77</u>:5500-5507.

19. Peusner, L. (1983) Electrical network representation of n-dimensional chemical manifolds, in <u>Chemical applications of topology and graph theory</u> (R.B. King,ed), Elsevier, Amsterdam.

20. Peusner, L. (1986) <u>Studies in network thermodynamics</u>, Elsevier, Amsterdam.

21. Thakker, K.M. and D.C. Mikulecky. Modeling and Simulation of the Nonlinear Dose-Plasma Concentration Response of Phenylbutazone on Periodic Multiple Oral Dosing: A Mechanistic Approach. Mathematical Modelling, 7:1181-1186 (1986).

22. Westerhoff, H. V. and K. van Dam (1987) Thermodynamics and Control of Biological Free-energy Transduction, Elsevier, Amsterdam.

23. White, J.C.(1979) Reversal of methotrexate binding to dihydrofolate reductase by dihydrofolate : Studies with pure enzyme and computer modeling using network thermodynamics, J. Biol. Chem. 254:10889-10895.

24. White, J.C. and Mikulecky D.C. (1981) Application of network thermodynamics to the computer modeling of the pharmacology of anticancer agents: A network model for methotrexate action as a comprehensive example, Pharmacol. Ther. 15: 251-291.

25. Wyatt, J.L., D. C. Mikulecky and J. A. DeSimone (1980) Network modeling of reaction-diffusion systems and their numerical solutions using SPICE, Chem. Eng. Sci. 35: 2115-2128.

26. Rashevsky, N. (1954) Topology and life:In search of general mathematical principles in biology and sociology, Bull. Math. Biophys. 16:317- 349.

27. Rosen, R. (1985) Information and complexity, in Ecosystem theory for biological oceanography, Canadian Ball. Fish. and Aquat. Sci. 213:221-233.

28. Rosen, R. (1985) Organisms as casual systems which are not mechanisms: An essay into the nature of complexity, in Theoretical Biology and Complexity, R. Rosen, ed. Academic Press, N.Y.

29. Rosen, R. (1985)   Anticipatory systems,
Pergamon, London.

30. "Some of the chemical reactions occurring in
the cell." fig. 2-35 in Albers, B., D. Bray, J.
Lewis, M. Raff (1983) <u>Molecular Biology of the
Cell,</u> Garland Publishing, Inc., N. Y. pp 82-83.

31. Mikulecky, D. C. (1990) Modeling Intestinal
Absorption and Other Nutrition-Related
Processes Using PSPICE and STELLA, <u>J. Ped.
Gastroent. and Nut.</u>, in press.

32. Mason, S.J. and H.J. Zimmermann (1960)
<u>Electronic circuits, signals, and systems</u>,
Wiley, N.Y.

33. McCulloch, W. S. (1988) <u>Embodiments of Mind</u>,
M.I.T. Press, Cambridge, MA.

34. Hopfield, J. J. (1982) Neural networks and
physical systems with emergent collective
computational abilities, <u>Proc. Natl. Acad. Sci.
USA 79</u>:2554-2558.

35. McClelland, J. L., D. E. Rumelhart, and the PDP
Research Group (1986) <u>Parallel Distributed
Processing: Explorations in the microstructure
of cognition, Volume 2, Psychological and
biological models</u>, M.I.T.Press, Cambridge, MA.

36. Grossberg, S. (1988) <u>Neural networks and
natural intelligence</u>, M.I.T. Press,
Cambridge,MA.

37. Albers, B., D. Bray, J. Lewis, M. Raff (1983)
<u>Molecular Biology of the Cell,</u> Garland
Publishing, Inc., N. Y.

TRAVELING WAVES IN EXCITABLE MEDIA

John J. Tyson

Department of Biology
Virginia Polytechnic Institute and State University
Blacksburg, Virginia 24061

Electrical stimuli travel along nerve-cell axons, phalanxes of amoebae crawl over a rotting leaf in response to a chemoattractant, a flame front spreads through a fuel mixture, waves of contraction pass rhythmically through heart muscle, regions of star formation move majestically through interstellar space at the leading edge of the arms of spiral galaxies, waves of depressed brain-cell activity rotate over the cortex of a migraine victim. These are all examples, more-or-less well-documented, of excitable media through which travel organized, self-sustaining waves of "excitation". What features do these diverse systems have in common? Usually they are characterized by a resting state which can be stimulated to undergo a rapid change of state. This local change of state triggers neighboring regions to undergo a similar transition, and thus the excitation propagates through space. Behind the wave of excitation, the medium typically recovers to its stable resting state in preparation for further stimulation. For instance, the cool unreactive mixture of fuel and oxidizer can be triggered by a spark to undergo an explosive chemical reaction that propagates rapidly through the mixture. If the reaction is self-extinguishing before much fuel and oxidizer is consumed, then the mixture will cool down and return to the resting state, ready to propagate another wave of oxidation in response to another spark.

Excitable media of this type can often be described adequately by a simple mathematical model

$$\frac{\partial u}{\partial t} = \nabla^2 u + f(u,v) \tag{1a}$$

$$\frac{\partial v}{\partial t} = \delta \nabla^2 v + \varepsilon g(u,v) \tag{1b}$$

In this system u and v are state variables, functions of time and space (one-, two-, or three-dimensional). ∇^2 is Laplace's operator ($\partial^2/\partial x^2 + \partial^2/\partial y^2 + \partial^2/\partial z^2$ in three dimensions), and δ is the ratio of "diffusion" coefficients of u and v. Epsilon is a small parameter, $0<\varepsilon<<1$, reflecting the fact that v characteristically changes on a much slower time scale than u. We refer to the fast-variable, u, as the "excitation" variable and the slow variable, v, as the recovery variable. The local dynamics of u and v, given by the pair of ODEs $\dot{u} = f(u,v)$, $\dot{v} = \varepsilon g(u,v)$, is described by the phase plane portrait in Fig. 1. The resting state (u^*, v^*) is given by the unique solution of $f(u^*, v^*) = g(u^*, v^*) = 0$.

To the extent that natural excitable media can realistically be described by this mathematical caricature, Eq. (1) and Fig. 1, they share many common features. In one spatial dimension, a solitary pulse of excitation travels at a characteristic speed through resting medium, and periodic pulses travel at a speed which (generally) decreases as the spacing between waves decreases. In two spatial dimensions, there are two characteristic patterns of periodic wave

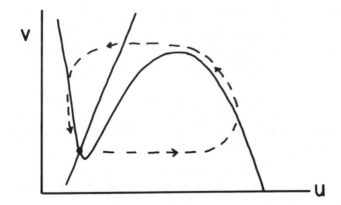

Fig. 1.   Typical phase plane portrait for an excitable medium.   Solid lines are the nullclines, $f(u, v) = 0$ and $g(u, v) = 0$.   Dashed line is a typical trajectory initiated by a suprathreshold stimulus.

propagation: expanding concentric circular waves (target patterns) and rotating spiral waves. In three dimensions, rotating spiral waves generalize to scroll-shaped waves rotating around a one-dimensional filament of inactivity. As the scroll rotates, the filament slowly moves through space according to rules which are as yet inadequately described and understood.

The mathematical caricature has been studied by many techniques: brute-force numerical methods, bifurcation theory, singular perturbation analysis, and cellular automaton models. Certain crucial features have emerged from these studies. The properties of wave propagation in one spatial dimension are profoundly influenced by the "dispersion" relation, which specifies the wave speed c as a function of the time elapsed T since the previous wave front: $c = \sigma(T)$. To describe wave propagation in two spatial dimensions, in addition to time intervals, one must also keep track of the curvature of the wave front, because the rule for wave front motion is

$$N = c + \varepsilon K \tag{2}$$

where N = normal velocity, c = plane-wave speed = $\sigma(T)$, and K = wave-front curvature. In three spatial dimensions, further complications arise: the motion of the filament is described by equations of the form

$$V_n = b_2\kappa + c_2w_s - a_2w^2 + O(\kappa^2, w^4, etc.) \tag{3a}$$

$$V_b = c_3\kappa + c_4w_s - a_3w^2 + O(\kappa^2, w^4, etc.) \tag{3b}$$

where V_n and V_b are the normal and binormal velocities of the filament in the local Frenet coordinate system, s is arclength along the filament, $\kappa(s)$ is the curvature of the filament, and $w(s)$ is the local twist rate of the scroll wave around the filament. The coefficients, a_i, b_i, c_i , can only be determined phenomenologically, at present. The adequacy of Eq. (3) breaks down when two filament segments approach each other too closely in space, and there is presently no theory to describe such filament interactions.

There are four examples of excitable media for which the interchange of theory and experiment has been especially fruitful. (1) The nerve axon is a perfect example of a one-dimensional excitable medium. The electrophysiology of neural membranes is known in great detail, and the kinematics of nerve impulse propagation can be derived elegantly from the dispersion relation, $c = \sigma(T)$. (2) The aggregation of slime mold amoebae into a multicellular slug is a fascinating example of a two-dimensional excitable medium. The basic biochemistry of the chemical signalling system has recently been uncovered, and a quantitative theory of cell-to-cell signalling and aggregation is now possible. (3) The Belousov-Zhabotinsky reaction is a convenient chemical system for studying wave propagation experimentally (the reaction can easily be constituted and observed in one, two or three spatial dimensions) and theoretically (the mechanism of the reaction is thoroughly understood). The BZ reaction is especially important for testing the agreement between theory and experiment. (4) The thick muscular wall of the heart ventricle is a three-dimensional excitable medium with two alternative modes of action: the normal heartbeat, in which waves of neuromuscular activity spread smoothly and controllably from one wall of the ventricle to the other walls, and ventricular flutter, in which scroll-shaped waves of activity rotate rapidly and uncontrollably around slowly moving filaments of inactivity. Because ventricular flutter is often the prelude to cardiac arrest, it is especially important to understand how flutter gets started (how normal wave propagation is rearranged to a scroll-shaped wave), how it develops (how the filament moves through the heart muscle as the scroll wave rotates), and how it might terminate either naturally or by medical intervention (how the scroll wave might annihilate itself or be annihilated).

For further information, readers should consult the following sources.

1. Nerve impulse propagation
    a. H. C. Tuckwell, *Introduction to Theoretical Neurobiology: Volume 2*, Cambridge Univ. Press, Cambridge, 1988.
    b. J. Rinzel, in *Dynamics and Modelling of Reactive Systems* (edit. by W. E. Stewart, W. H. Ray and C. C. Conley), Academic Press, New York, 1980.
2. Slime mold aggregation
    a. J. L. Martiel and A. Goldbeter, *Biophys J.* 52:807-828 (1988).
    b. J. J. Tyson, K. A. Alexander, V. S. Manoranjan and J. D. Murray, *Physica D* 34:193-207 (1989).
3. Belousov-Zhabotinsky reaction
    a. R. J. Field and M. Burger, eds., *Oscillations and Traveling Waves in Chemical Systems*, Wiley-Interscience, New York, 1985.

    b.   J. P. Keener and J. J. Tyson, *Physica D* 21:307-324 (1986).

4. Cardiac muscle

    a.   A. T. Winfree, *When Time Breaks Down: The Three-dimensional Dynamics of Electrochemical Waves and Cardiac Arrhythmias*, Princeton Univ. Press, Princeton, 1987.

    b.   V. S. Zykov, *Simulation of Wave Processes in Excitable Media*, Manchester Univ. Press, Manchester, 1987.

5. General theory

    a.   J. J. Tyson and J. P. Keener, *Physica D* 32:327-361 (1988).

    b.   J. P. Keener, *Physica D* 31:269-276 (1988).

    c.   P. K. Brazhnik, V. A. Davydov and A. S. Mikhailov, *Theor. Math. Phys.* 74:300-306 (1988).

6. Numerical studies

    a.   A. T. Winfree, *SIAM Rev* 32:1-53 (1990).

    b.   M. Gerhardt, H. Schuster and J. J. Tyson, *Science* 247:1562-1566 (1990).

VESTIBULAR HAIR CELL TRANSDUCTION IN THE STRIOLAR REGION OF THE OTOLITH ORGAN

J. Wallace Grant

Virginia Polytechnic institute and State University
Engineering Science and Mechanics Department
Blacksburg, Virginia   24060-0219

The Otolith Organs are the mammalian linear motion sensing system. When these are combined with the semicircular canals, which sense rotational motion, they form the vestibular system. The vestibular system is located in non-auditory portion of the inner ear. The vestibular system and the auditory system both contain hair cells which transduce the basic mechanical signals into nervous system action potentials. The mechanism of transduction in the hair cells appears to be common throughout the inner ear.

The striolar region of the otolith organs is a special region of transduction. This region is located in the center portion of these flat plane type of structures. The peripheral regions of these organs consist of a gel layer and an otoconial layer which is attached to the gel layer. The otoconial layer is composed of calcium carbonate crystals making it a much denser than any other material in the inner ear. The gel layer is attached to a sensory cell base which is rigidly attached to the skull.

With any acceleration of the skull or tilt with respect to the gravity vector the otoconial layer shears the gel layer. This shearing of the gel layer constitutes the fundamental mechanical sensory mechanism of the otolith organs. The gel layer and otoconial layer form an overdamped second order system whose transfer function is flat below its lower corner frequency. This transfer function relates otoconial layer displacement to the sum of skull acceleration and the acceleration of gravity. No sensor can detect the difference between gravity and inertial acceleration.

The hair cells project small cilia upward from their cell bodies into the gel layer. The cilia are bent with any shear deformation of the gel layer. These cilia are of two types: many shorter steriocilia and a single kinocilia with each hair cell. When the steriocilia are bent, deforming their upper tip, ion channels are opened which result in depolarization of the cell. This depolarization modulates the action potential firing frequency of

the nerve associated with the hair cell. The
kinocilia appear to perform only a stiffening
function in the transduction process. The kinocilia
contain a 9+2 microtubual structure associated with
motility of cilia in general. In the kinocilia this
microtubular structure lacks part of the necessary
protein components for fast motility and for this
reason probably only forms a stiffening function.
This stiffness is necessary for transduction in the
striolar region.

The cilia of the hair cells form a unique
arrangement. The steriocilia of which there are
approximately 75 to 100 per cell form an organ
pipe arrangement of gradually increasing height
from front to back. The kinocilia is located at the
very back and is a least 10 times taller than the
tallest steriocilia in the striolar region. The
steriocilia and single kinocilium are connected at
their tips with protein tethers which tie the
steriocilia together structurally and to the base of
the kinocilia. In the peripheral region of the otolith
the kinocilia are attached to the gel layer. In the
striolar region the gel layer is absent and this
space is filled with a fluid called endolymph which is
also located above the otoconial layer. The
steriocilia and kinocilia project from the hair cell
base into this fluid in the striolar region. With any
motion of the gel layer wall, which forms the outer
boundary of the striolar region, fluid motion which

tracks the gel wall motion is induced. This induced fluid motion results in flow around the kinocilia and the drag force produced by this flow results in bending of the stiff kinocilia. The bending of the kinocilia then produces a normal transduction of the hair cell. Thus the hair cell transduction in the striolar region is fluid coupled instead of the normal coupling to the gel layer.

If the drag force imposed on the kinocilia is directly proportional to the fluid velocity in the striolar region (this fact will be discussed later) then the hair cell signal in this region should be equal to the first time derivative of the signal in the peripheral region. This is a consequence of the fluid velocity in this region being the first derivative of the gel wall displacement. This fact has been somewhat confirmed with nerve fiber recordings from the peripheral and striolar region. The peripheral region contains signals which are tonic and the striolar region contains signals which are phasic. Here, tonic refers to nerve frequency which is proportional to the acceleration input and phasic refers to a signal which is proportional to the time rate of change of the acceleration input. Thus the peripheral region of the otolith reports acceleration information and the striolar region reports time rate of change or jerk information to the central nervous system.

The fluid transduction mechanics in the striolar region has only recently been investigated. In fact the entire transduction process in the otolith organs has been studied very little and for that reason the knowledge about this organ is rather sparse. Even less studied is the striolar region. In order to investigate the fluid coupling in the striolar region two concepts must be developed and understood. These include the fluid drag acting on the kinocilia and the fluid velocity.

The fluid drag can be calculated using linearized creeping motion equations and the solution developed by Lamb and Oseen. This solution is applicable for Reynolds numbers less than 10 for smooth cylinders. The flow in the striolar region of the otolith has a Reynolds number in the range from 10^{-8} to 10^{-1}, thus this solution is applicable. The drag force produced per unit length **w(y)** acting on the kinocilia is given by

$$\mathbf{w(y)} = \frac{4\pi}{\left(\ln\frac{8}{R_e} - 0.0772\right)} \, \mu \, v$$

where v=fluid velocity, μ=fluid viscosity, and Re=Reynolds number. This equation can be simplified to the following over the range of Reynolds numbers encountered

$$\mathbf{w(y)} = \mu v$$

This approximation introduces only a minor error in the range of Reynolds numbers of interest, but large errors are possible in certain ranges. This approximation allow for a simple solution. A more extensive solution without the approximation can only be had by numeric integration.

The fluid velocity **v** represents the fluid velocity measured with respect to the kinocilium. Since the kinocilium can also be in motion due to its stiffness and bending the velocity **v** can come from the fluid motion induced by the striolar walls or from the kinocilium moving through the fluid which is at rest. Taking these two sources into account the velocity becomes

$$\mathbf{v} = \frac{y}{b} V - \frac{du}{dt}$$

where y=fluid depth, b=height of the striolar region V=velocity of the otoconial layer, u=the kinocilium displacement, and t=time.

When this expression is combined with the drag force equation it can be integrated utilizing the beam equations to achieve an overall governing equation of kinocilia motion. The resulting integrated equation is a first order differential equation describing the motion of the kinocilia in the striolar region. This equation is

$$\frac{d u_z}{dt} + 8\frac{EI}{z^+\mu}u_z = 0.733\frac{\ell}{b}V$$

where u_z=the displacement of the end of the
kinocilium, EI=effective bending stiffness of the
kinociliua, and l=kinocilium height. This simple
first order differential equation is an
approximation to the motion dynamics of the
kinocilium.

The time constant for the kinocilium to return
to its equilibrium position after an initial deflection
can be calculated with this equation. This time
constant is about 3 seconds when values for the
physical constants in the equation are utilized.
Time constants of this magnitude show up in time
recordings of the nervous signals recorded from
the nerve fibers originating in the striolar region.
This is some confirmation that the approximate
analysis is correct. When this experimental data is
displayed as a function of inertial excitation
frequency it indicates that the response is not
linear as suggested by the differential equation.
This result would indicate that the linearization
step utilized in the drag equation must be included
in order to completely characterize the dynamic
response of the system.

The force reaction at the base of the
kinocilium can also be calculated using simple beam
equations. This analysis shows that the force
reaction generated at the base of the kinocilium is
approximately in the range from 1 to 600 pN

depending upon the inertial motion of the skull. Measured values of the steriocilia tuft stiffness confirms this range of force values. These measured values of tuft stiffness range from 600 to 130 pN/micro m. These values produce maximum deflection ranges from one to five micrometers. This is the range of maximum deflections is within the range of motion expected motion of the steriocilia tufts.

With this confirmation of the linearized models ability to predict realistic physiologic values the nonlinear effects of the drag force should now be included. These effects will come closer to matching the recorded experimental response. With confirmation of the mechanical system response the hair cell contribution to the overall response can be evaluated.

ACKNOWLEDGEMENT

This research was sponsored by the Naval Medical Research and Development Command under work unit 61153N MOR4106.008-7008. The views expressed in this article are those of the author and do not reflect the official policy or position of the Department of the Navy, Department of Defense, nor the U. S. Government.

III.

BIOTECHNOLOGY

THE BIOLISTIC PROCESS -

AN EMMERGING TOOL FOR RESEARCH AND CLINICAL APPLICATIONS

J. C. Sanford
Cornell University
Hedrick Hall
Geneva, NY 14456

BACKGROUND

The biolistic process is a unique mechanism which employs high velocity microprojectiles to deliver substances into cells and tissues. This new process has been called by different names. It has been referred to as the particle gun method, the microprojectile method, the gene gun method, the particle acceleration method, the bio-blaster method, etc. The inventors of the process have coined the term "biolistic" to describe both the process and associated apparatus. Biolistics stands for "biological ballistics" and seems most appropriate -- given that biological materials are being shot into biological targets.

The biolistic process was first developed by myself, in collaboration with Ed Wolf and Nelson Allen (1). Much of the critical early work was done by Ted Klein in my lab (1-10). Later key developments were accomplished through collaborations between my lab and the labs of Johnston (5,11,12), Wu (2,7,8), Boynton, Gilham, and Harris (6), Tomes and Weissinger (3), Fromm (4,8,9,10), Bogorad (13), Daniell (14), and others (15-17). Important later advances have also come from scientists at Agracetus, Inc. (18-20). These early developments have all been reviewed previously (21,22). In one of these reviews (21), certain key developments were anticipated. Specifically, it was considered important that if the biolistic process was to find truely wide-spread utility, we would need to see the demonstration of biolistic transformation of bacteria (procaryotes - the smallest of targets), fertile corn (the 'Holy Grail' for plant genetic engineers), and intact animals. In the relatively short time since that review, each of these major tests has been met. We can now routinely achieve high efficiency biolistic transformation of the hard-to-transform gram positive bacterium *Bacillus megaterium* (23), and the always-present gram negative *Escherichia coli* (24). At least two independent groups report the production of biolistically-produced fertile transgenic corn plants (25). Perhaps most importantly, and (for some) most suprisingly, genetic material has effectively been introduced into intact animals (11). Other important developments since the last review include pollen transformation (26), Gus expression in higher plant chloroplasts (27), and transformation of diverse animal cells *in vitro* (unpublished, and 28,29).

APPLICATIONS

The biolistic process provides unique capabilities in terms of delivery of substances into intact cells and tissues. Certain important genetic "targets" can now be reproducibly transformed, which were not transformable by any other method. The process is unusually simple and rapid, and appears to be applicable for nearly any type of target cell or tissue, in any type of organism.

Gene transfer methods have already been developed for many important species of life in all kingdoms. Certain bacterial, fungal, plant, and animal cells are being routinely transformed in labs all over the world. This might lead a casual observer to believe that a new method for gene transfer would not be needed. However, despite the existence of diverse transformation techniques, gene delivery remains a limiting step in innumerable fields of research and development. Even where there are published reports of gene transfer into a given target genome, scientists may find such methods unworkable. A given method may be unreproducible, or many do not have the necessary delivery efficiency, or it may not be fast enough, or cheap enough. Alternatively, highly effective methods may exist which require infectious agents, but these present related complications -- biological and legal.

The biolistic process has been proven to be effective in a wide range of targets; including procaryotes and eucaryotes / microbes, animals, and essentially any higher plant / single cells, tissues, and intact organisms. Given these demonstrations, the future utility of the process which can be foreseen to include: 1. basic transformation studies for research purposes; 2. genetic engineering of new and useful plants and animals; 3. somatic gene delivery in animals and man; 4. genetic vaccination and/or delivery of pharmaceuticals.

Basic Transformation Studies -

Approximately 100 labs are now using biolistic technology to conduct gene transfer experiments on various organisms. This includes basic gene delivery studies on mitochondria, chloroplasts, hard-to-transform primary animal cell lines, bacteria, fungi, fish, insects, bull sperm, nematodes, plant cell suspensions, plant embryos, meristems, pollen, various plant parts, intact frogs, and intact mice (including skin, ear, muscle, and liver). In the hard-to-transform systems where the biolistic process has proven effective, the use of this technology is becoming a routine research procedure,   i.e. for genetic studies of mitochondria, chloroplasts, and corn cell suspensions. It is reasonable to expect that the use of the biolistic technique as a research tool will continue to grow rapidly.

The use of the biolistic process as a screening tool for gene expression in differentiated tissues has become especially well established, and has been reviewed by Klein et al. (30,31). The value of the biolistic process for this purpose is obvious. Numerous genetic constructs coding for different gene products, or a single reporter gene having different promoter configurations, can be delivered into any specific tissue to rapidly determine what effect different constructs have on different tissues.   No other gene delivery technology can provide this important capability.

Genetic Engineering of New and Useful Strains of Organisms-

A large amount of effort is now being put into the production of transgenic organisms which might have value to man.   In most species, production of transgenics is still difficult or impossible.   For commercial utilization of domesticated organisms, production must not only be possible, it must be efficient.

Production of transgenic plants is feasible by several methods, but the only methods which appears commercially promising for our most important crops (i.e. grains, soybeans) appears to be the biolistic process. In fact the limited host-range of *Agrobacterium*, and the extreme difficulty of regenerating most plants from protoplasts, makes biolistics the only promising candidate as a "universal" plant transformation methodology.

Production of transgenic animals can be achieved in certain species by micro-injection of surgically explanted egg cells.   While this method has proven very valuable in mice, it remains difficult and is not easily applied to larger domesticated animals.   However, preliminary experiments with bull sperm show that sperm cells can be biolistically penetrated by microprojectiles, penetrated cells can be physically separated on the basis of density, and the resulting cells can remain viable (32).   This raises the interesting possibility that sperm could be bombarded *in vitro*, and could then be used to deliver genetic material to the zygote *in utero*.   Such methodology, if feasible, might easily be extended to a very wide range of animals.   Likewise, because the biolistic process is effective on intact tissues, ovaries and testis might be bombarded directly, resulting in transgenic gametogenic tissue.   Eggs or zygotes might also be bombarded directly, although it is not clear that this would provide any clear advantage over microinjection.

Somatic Gene Therapy -

There are methods involving retroviruses which allow effective somatic gene therapy in experimental animals such as mice.   In particular, hematopoietic cells of the bone marrow have been extracted, transformed with retroviral vectors, and injected back into live animals.   Under the proper regimen, such cells have re-established in the bone marrow and have provided an effective cure for certain genetic defects.   However, this approach to gene therapy has some disadvantages.   Firstly, it involves an infectious agent which is genetically related to viruses lethal to man.   This may create regulatory delays and biological complications. Secondly, this methodology is generally only effective on cells *in vitro*, and is therefore not amendable to most intact tissue types, since most cell types can not be easily cultured *in vitro* and then reintroduced into the body. The biolistic process offers a unique mechanism for transforming intact somatic tissue *in situ*. What practical or clinical value would this capability provide? People are just beginning to explore this question.

Experiments thus far indicate that genetic material can be biolistically introduced and expressed within intact epidermis, dermis, and liver, *in situ* (11). Likewise, genes have been biolistically transformed into various differentiated primary cell types, including myotubes (muscle cells)(29), and other hard to transform cell lines (unpublished).

The most effective biolistic delivery into intact animal tissues has been through bombardment of skin. The prospective beauty of *in situ* skin transformation would be that genes could be introduced in a non-invasive manner using a procedure that might be no more traumatic than a simple vaccination. The expression of introduced genes would be limited to the area of skin treated. This might have value where a localized effect was desired as might be desired for various skin disorders. Alternatively the dermis is sufficiently vascularized such that by treating a sizeable area, skin transformation might result in significant circulating levels with the blood stream of secreted gene products such as insulin or growth hormone. Likewise, transient biolistic skin transformation might prove ideal for genetic vaccines, which are described later.

The value of biolistic muscle transformation would be the extensive surface areas available for bombardment through relatively superficial surgical procedures. This tissue is also highly regenerative and is very tough - minimizing concern for tissue damage from bombardment. Muscle is highly vascularized, making it preferable to skin where higher circulating levels of a secreted gene product may be desired.

Preliminary experiments indicate that liver (and presumeably other soft tissues such as spleen, intestine, pancreas) can be biolistically transformed *in situ*. While these results are very preliminary, it may indicate that exciting alternatives for gene therapy are feasible. Such soft tissues can only be exposed through major surgery, however, numerous genetic conditions would justify such surgery. The liver is the site of numerous genetic defects which are already recognized as candidates for gene therapy. The pancreas might be treated biolistically to re-establish its normal insulin-producing function. The spleen, like the bone marrow, contains hematopoietic cells, and might be treated biolistically to address some of the blood disorders most commonly considered for gene therapy. Many questions will need to be answered before we will know if such applications are actually feasible.

Genetic Vaccination/Delivery of Pharmaceuticals -

While most clinical applications of the biolistic process require stable genetic transformation of cells within tissues, there are some applications where transient gene expression is actually preferrable. There are other applications where it is non-genetic material that would be delivered biolistically.

One of the most exciting prospective applications for the biolistic process is to create "genetic vaccines". Effective vaccines have been developed for many important diseases, but for many diseases vaccines have not proven effective. One reason a vaccine fails to be effective is because the body's immune reponse to the relevant antigens, while detectable, is simply not strong enough. This is typically because the antigen is not present in sufficient quantity, or for a sufficiently long time, in the blood stream (hence booster shots sometimes help). If dermis cells were bombarded such that they transiently expressed and secreted a foreign antigen, they could produce relatively large amounts of the antigen over a period of 5-15 days. This should allow us to create a much stronger antigenic response than through conventional vaccination. The antigenic response should eventually lead to a cytotoxic response against the expressing cells, clearing the body of all secreting cells, making the process truly transient - very distinct from "gene therapy" as we normally think of it.

Genetic vaccination is likely to be the first clinical use of the biolistic process and when proven valid, it is likely to come into use relatively soon. Its use will probably begin with genetic vaccination of animals for antibody production for the research community, and will then be extended to use on domestic animals, prior to human clinical use.

Not all biolistic applications will deliver genetic material. For example, it should be possible to deliver pharmaceuticals into intact tissue by the biolistic process. Many pharmaceutical uses are limited by the fact that a given pharmaceutical fails to cross certain barriers, such the epidermis. Therefore topical treatment of many localized skin disorders are ineffective simply because the pharmaceutical fails to penetrate the outer layers of the skin. Likewise, it is often desireable to have a pharmaceutical which is persistently released, or is localized, or both. In theory, the biolistic process should be effective in delivering pharaceuticals into the epidermis and/or the dermis in "slow-release" particles 5-100 microns in size. There should be numerous advantages to this, but there has been no work in this area up to this time.

TECHNOLOGY ADVANCEMENT

While there apppear to be very exciting and diverse potential applications for the biolistic process, there remain many technical problems and much work remains to be done. Developmental work is really just beginning. Uncertainties remain and certain prospective applications remain only theoretical at this point. In order to advance biolistic technology, further work is needed in 3 basic areas: 1. Improved accelerator designs; 2. Improved microprojectiles with improved coating and dispersion methods; 3. better understanding of the biological

determinants of cell penetration and survival with detailed studies
optimizating the process in each   application area.

Improved accelerator designs -

        No fundamentally new microprojectile acceleration mechanisms have
been demonstrated, which were not outlined in the first biolistic
publication (1).   However there have been numerous creative
modifications and embellishments in particle accelerator designs (18,
33,34,35).   The particle acceleration system which has been most
extensively tested and most widely utilized is the commercially available
PDS-1000 (DuPont).   Up to this time, there has been no direct evidence that
any of the alternative designs are any more effective than the PDS-1000.

    While the PDS-1000 system has certain proven capabilities, it has certain
important limitations including: relatively fixed mean initial velocity,
potentially severe acoustic shock to cells, non-uniform dispersion of
microprojectiles, shot-to-shot variation, and inappropriate configuration
for bombardment of intact animals.   The PDS -1000 system has been
modified with a "mousetrap" attachment which allows bombardment of
intact mice (11), but this mechanism is not suitable for larger animals.

        For the reasons given above, my lab has, in collaboration with others,
developed a series of improved biolistic devices which represent a major
improvement over the PDS-1000 (36).   These new devices employ a high
pressure helium gas shock wave to accelerate miroprojectiles, using
interchangeable launch mechanisms.   We now have two basic designs: 1. a
bench-mounted vacuum chamber system suitable for bombarding cells in a
petri plate; and   2. a "wand" configuration suitable for use as a hand-held
tool for surgical and dermal experiments on intact animals of any size.   I
believe   the bench-mounted system may soon become available
commercially as an upgrade of the PDS-1000 unit.   I believe "wands" may
soon become available as a dedicated system for intact animal experiments.

        The bench-top system has been shown to be markedly superior to our
previous designs.   Power level can be controlled, acoustic shock can be
minimized, dispersion over a large bombardment surface is excellent,
repeatability is improved, and the system is extremely versatile for
different applications.   The "wand" combines these advantages with its
dedicated design for use with intact animals of any size.   The new systems
have been directly compared to the PDS-1000 system using *Bacillus
megaterium*, yeast, tobacco cell suspensions, animal cells *in vitro*, and
intact mouse tissues.   In every case a dramatic increase in transformation
rates was seen, with rates increasing 4-100 fold over the PDS-1000 rates
(36).   In addition, transformants appeared more uniformly over the area of
bombardment, with no obvious zone of death or cratering -   problems
commonly seen with the PDS-1000.   I believe that with further refinement
these new helium systems will be the basis for a new generation of

apparatus which will be widely utilized for many years, reliably giving very high rates of transformation is extremely diverse biological systems.

Improved Microprojectiles/Improved DNA Coating and Dispersion -

The microprojectiles presently in use are either tungsten particles from Sylvania, or gold particles from Alfa. These particles are far from ideal. The tungsten particles come in numerous mean size ranges but are highly irregular in shape and each lot is extremely heterogeneous in size distribution. These factors contribute to severe clumping upon DNA precipitation, and prevent us from identifying optimal particles sizes for each application. It is assumed that the vast majority of the particles in these mixtures are the wrong size for any given application. Larger particles kill target cells, smaller particles fail to penetrate. In addition, we have found that tungsten is toxic to numerous cell types, including some of the cells we are using regulary to monitor efficiency, such as tobacco cell suspensions and *E. coli*. Currently available gold particles are more uniform in terms of size and shape than tungsten, but are not available in the various size ranges which would allow systematic optimization for diverse applications. It is hoped that uniform gold particles will soon become available in numerous discrete sizes falling in the range of 0.1- 6.0 microns.

Present methods precipitate DNA onto the outer surface of the microprojectiles. Such coating is non-uniform and contributes to clumping of particles. After considerable effort we have not been able to make any fundamental improvements in the coating process. However, we have found that uniformity can be increased and clumping can be decreased by continuous agitation during and between all pipetting steps of the commonly used protocol - from the aliquoting of the stock tungsten suspension, to addition of spermidine, to the loading of macroprojectiles.

There is reason to believe that the surface molecules of the microprojectiles are exposed to extreme heating during flight due to high velocity impact with residual gas molecules. While this heat pulse is very short lived and superficial in nature, it may remove or degrade much of the DNA coating, especially at higher velocities. Therefore methods which would encapsulate the DNA, or would encapsulate the entire DNA-coated microprojectile may be highly desirable, especially for applications requiring very high velocity. Litttle work has been done in this area thus far.

Understanding Biological Determinants -

Even when all the physical parameters of the biolistic process have been optimized (i.e. particle composition, size, coating, velocity, etc.), numerous biological variables will need to be understood and optimized before the process will be maximally effective. These biological factors include cell size, target tolerance of vacuum, presentation (i.e. orientation, spread) of cells or tissue for bombardment, cell turgor, etc. Of special importance is the physiological status of the target. We know that cell culture stage, mitotic stage, and general cellular health are important factors affecting biolistic efficiency. As might be expected, cell turgor is important, and addition of high concentrations of osmoticum to the cellular medium prior to bombardment is crucial for many applications. It is clear that some cells tolerate a severe acoustic shock better than others, and this survival interacts with the level of osmoticum used.

Optimizing more than several variables can be very time consuming, especially where variables interact. Because there are multiple physical and biological parameters which need to be optimized for each application area, a series of fractional factorials seems to be a preferred experimental approach, clustering within seperate experiments factors expected to interact. In this way a data base should rapidly develop which will let us predict optimal experimental conditions for new targets.

REFERENCES

1. Sanford, J. C., T. M. Klein, E. D. Wolf, and N. Allen. 1987. Delivery of substances into cells and tissues using a particle bombardment process. J. Part.Sci. and Tech. 5:27-37.
2. Klein, T. M., E. D. Wolf, R. Wu, and J. C. Sanford. 1987. High-velocity microprojectiles for delivering nucleic acids into living cells. Nature 327:70-73.
3. Klein, T. M., M. E. Fromm, A. Weissinger, D. Tomes, S. Schaaf, M. Sleeten, and J. C. Sanford. 1988. Transfer of foreign genes into intact maize cells using high velocity microprojectiles. Proc. Natl. Acad. Sci. 85:4305-4309.
4. Klein, T. M., M. E. Fromm, T. Gradziel, and J. C. Sanford. 1988. Factors influencing gene delivery into Zea mays cells by high-velocity microprojectiles. Biotechnology 6:559-563.
5. Armaleo, D., G. Ye, T.M. Klein, K.B. Shark, J.C. Sanford, and S.A. Johnston. 1990. Biolistic nuclear transformation of Saccharomyces cerevisiae and other fungi. Current Genetics 17:97-103.
6. Boynton, J. E., N. W. Gillham, E. H. Harris, J. P. Hosler, A. M. Johnson, A. R. Jones, B. L. Randolph-Anderson, D. Robertson, T. M. Klein, K. Shark, J. C. Sanford. 1988. Chloroplast transformation of Chlamydomonas using high velocity microprojectiles. Science 240:1534-1538.
7. Cao, J., Y-C. Wang, T.M. Klein, J. C. Sanford, and R. Wu. 1990. Transformation of rice and maize using the biolistic process. In: Plant Gene Transfer(1989 UCLA Symposium) Alan R. Liss, Inc. pp21-33.
8. Wang, Y. C., T. M. Klein, M. Fromm, J. Cao, J. C. Sanford, and R. Wu. 1988. Transformation of rice, wheat, and soybean by the particle bombardment method. Pl. Mol. Biol. 11:433-439.
9. Klein, T.M., E.C. Harper, Z. Svab, J.C. Sanford, M.E. Fromm, P. Maliga. 1988. Stable genetic transformation of intact Nicotiana cells by the particle bombardment process. PNAS 85:8502-8505.
10. Klein, T.M., L. Kornstein, J.C. Sanford, and M.E. Fromm. 1989. Genetic transformation of maize cells by particle bombardment. Plant Physiology 91: 440-444.
11. Johnston, S. A., S. Williams, M. Reidy, M. Devit and J.C. Sanford. 1990.

In situ biolistic transformation of mouse tissues. Science (submitted).

12. Johnston, S. A., P. Anziano, K. Shark, J. C. Sanford and R. Butow. 1988. Transformation of yeast mitochondria by bombardment of cells with microprojectiles. Science 240:1538-1541.

13. Blowers, A.D., L. Bogorad, K.B. Shark, G.N. Ye, and J.C. Sanford. 1989. Studies on Chlamydomonas chloroplast trasformation: foreign DNA can be stably maintained in the chromosome. The Plant Cell 1:123-132.

14. Daniell, H., J. Vivekananda, B.L.Nielsen, G.N. Ye, K.K. Tewari, and J.C. Sanford. 1990. Transient foreign gene expression in chloroplasts of cultured tobacco cells after biolistic delivery of chloroplast vectors. PNAS 87:88-92.

15. Fox, T. D., J. C. Sanford, and T. W. McMullin. 1988. Plasmids can stably transform yeast mitochondria totally lacking endogenous mtDNA. PNAS 85: 7288-7292.

16. Cummings, D.J., J.M. Domenico, and J.C. Sanford. 1989. Mitochondrial DNA from Podospora anserina: transformation to senescence via biolistic delivery of plasmids. In: Mol. and Cell Biol. UCLA Symposium - New Series. Alan R. Liss Inc. NY pp91-101.

17. Fitch, M.M., R.M. Manshardtl, D. Gonsalves, J.L. Slightom, H. Quemada, and J.C. Sanford. 1990. Stable transformation of papaya via microprojectile bombardment. Plant Cell Reports (submitted).

18. Christou, P., D. E. McCabe, and W. F. Swain.1988. Stable transformation of soybean callus by DNA-coated gold particles. Plant Physiology 87:671-674.

19. McCabe, D.E., B.J. Martinell, and P. Christou. 1988. Stable transformation of soybean (Glycine max) plants. Bio/Technology 87:923-926.

20. McCabe, D.E., W.F. Swain, B.J. Martinell. Pollen-mediated plant transformation. European Plant Patent Application No. 87310612.4. Published June 8, 1988. Bulletin 88/23.

21. Sanford, J. 1988. The biolistic process - a new concept in gene transfer and bioligical delivery. Trends in Biotechnology 6: 229-302.

22. Sanford, J. 1990. Biolistic plant transformation - a critical assessment. Physiologia Plantarum (in press).

23. Shark,K.B., F.D. Smith, P.R. Harpending, and J.C. Sanford. 1990. Biolistic transformation of a prokaryote, *Bacillus megaterium.* (in preparation).

24. Smith, F.D., R. Harpending, K.B. Shark, and J.C. Sanford. 1990. A simple and rapid new method for transforming E. coli utilizing the biolistic process (in preparation).

25. Annoucements made at the UCLA Symposium - Molecular Strategies for Crop Improvement, April, 1990. Abstract: Lemaux, P.G. et al. - Selection of stable transformants from maize suspension cultures using the herbicide Bialaphos, Journal of Cellular Biochemistry suppl.14E, p. 304.

26. Twell, D., T.M. Klein, M.E. Fromm, S. McCormick. 1989. Transient expression of chimeric genes delivered into pollen by microparticle bombardment. Plant Physiology (in press).

27. Ye, G.N., H. Daniell, J. Rasmussen, and J.C. Sanford. 1990. The Gus gene as a marker for chloroplast transformation in higher plants. EMBO (in preparation).

28. Zelenin, A.V., A.V. Titomirov, V.A. Kolesnikov. 1989. Genetic transformation of mouse cultured cells with the help of high velocity mechanical DNA injection. FEBS Letters 244: 65-67.

29. Wiliams, R.S. and S.A. Johnston. 1989. Transformation of differentiated myotubes using DNA-coated microprojectiles. Nature (submitted).
30. Klein, T.M., B.A. Roth, M.E. Fromm. 1989. Regulation of anthocyanin biosynthetic genes introduced into intact maize tissues by microprojectiles. PNAS (in press).
31. Klein, T.M., B.A. Roth, and M.E. Fromm. 1989. Advances in direct gene transfer into cereals. In: Genetic Engineering, Vol. 2, edited by Jane K. Setlow. Plenum Publishing Corporation.
32. Foote, R.H. and S.R. Hough. 1990. Effect of the Cornell particle gun on bull and rabbit spermatazoa. Proceedings of the Society for the Study of Reproduction, Knoxville, TN. (Abstract).
33. Morikawa, H., A. Iida and Y. Yamada. 1989. Transient expression of foreign genes in plant cells and tissues obtained by a simple biolistic device. Appl. Microbiol. Biotechnol. 31: 320-322.
34. Zumbrunn, G., M. Schneider, and J.-D. Rochaix. 1989. A simple particle gun for DNA-mediated cell transformation. Technique   1:204-216.
35. Oard, J.H., D.F. Paige, J.A. Simmonds, and T.M. Gradziel. 1990. Transient gene expression in maize, rice, and wheat cells using airgun apparatus. Plant Physiol. 92:334-339.
36. Sanford, J.C., M. DeVit, F. Smith, R. Harpending, J. Russell, M. Roy, G.N. Ye, X.J. Ye, and S.A. Johnston. 1990. An improved mechanism for biolistic particle acceleration (in preparation).

Cyclodextrins and Cyclodextrin Glucanotransferases

Bernard Y. Tao

Biochemical and Food Process Engineering
Dept. of Agricultural Engineering
Purdue University
West Lafayette, IN 47907

INTRODUCTION

One of the powerful driving forces behind biotechnology is the promise of tailor-made enzymes and proteins. Such materials would have "engineered" properties, such as enhanced stability, altered substrate specificity and novel catalytic capabilities. While this dream of "manufacturing" new biocatalysts has yet to be realized, significant progress has been achieved towards developing the fundamental structural-functional relationships required to achieve this goal. One promising family of enzymes which are being studied in this regard are the glucosylases, comprised of the amylases, glucosidases, sucrases and glucosyltransferases.

This overview presents recent developments in the area of cyclodextrin glucanotransferases (CGTase) and their product cyclodextrins (CD). A review of current CD/CGTase technology will be presented, with extension to relate its importance on elucidating general glucosylase biocatalyst structural and functional relationships. Comparision to known structures of other enzymes is presented. A brief discussion of current research in genetic mutation of this class of biocatalysts to elucidate structural/functional features is also included.

Cyclodextrins (CDs) were first observed nearly a century ago (1). First thought to be a crystalline form of starch, the unique cyclic structure and physical chemistry of these oligosaccharides have been elucidated through the contributions of Schardinger, Freudenburg, Cramer and French (2). In the past decade, CDs have seen great industrial and commercial interest due their ability to form complexes with a variety of molecules (3), act as organic enzyme mimics (4) and provide enantiomeric chromatographic separations (5). In the last 5 years, CD manufacturing plants have been constructed in West Germany, Hungary, France, the U. S. and Japan. This growth is anticipated to dramatically increase as these materials continue to gain approval for use in pharmaceuticals, foods and fine chemicals.

99

As the second century of cyclodextrin research begins, attention is being focused on the biochemistry of CDs and their biocatalysts, cyclodextrin glucanotransferases (CGTases). Recent advances in protein structure and genetics have provided the tools to probe and alter the biocatalytic structure of CGTases and begin to understand their relationships to other amylolytic glycosylases. Current efforts in modification and alteration of CGTase structure have begun to yield critical information needed to "engineer" biocatalyst activity. These results, combined with information about carbohydrate conformation and structure, will eventually lead to the "engineering" of new glycosylase biocatalysts with significant industrial importance for the food, chemical, environmental and energy industries.

STRUCTURE AND PROPERTIES OF CYCLODEXTRINS

Cyclodextrins (CDs) consist of six (α), seven (β) or eight (γ) glucose residues linked by α-1,4 bonds (Figure 1). CDs possess hydrophobic interiors and hydrophilic exteriors, due to the orientation of the glucose hydroxyl groups. This structure allows the formation of inclusion complexes with apolar compounds (6). Table I presents CD cavity size and solubility.

Table I. Properties of Cyclodextrins

CD type	α	β	γ
No. of glucoses	6	7	8
Cavity size (angstroms)	5.7	7.8	9.5
Solubility (gms/100 ml of water at 25 C)	14.5	1.8	23.2

The formation of chemically-modified cyclodextrins for immobilization, polymer creation and addition of specific ligands has also been noted (7, 8).

Stabilize and protection of sensitive host molecules, such as flavors, odors, or pharmaceuticals is the primary advantage of CD complexation. CD complexes sharply reduce the volatility, chemical, thermal and photo reactivity of guest molecules. Sensitive bioseparation techniques have been noted for CDs. For example, CDs can remove reactive components from fruit juices to prevent oxidation or eliminate bitterness. Chromatographic CD media provide chiral separation, selective component removal and modified chemical reactivity. A number of modified and polymerized CD materials have gained acceptance as separation media (9).

Chemically modified-CDs have also been used to mimic enzymatic reactions (10). Modified CDs have catalyzed the de-esterification, hydrolysis, selective substitution, addition and isomerization of a variety of molecules (9). Advances in the chemical synthesis of CDs (11) and unusual "manno" CDs (12) promise novel new CD-like materials in the future.

Many papers have been published describing the functionality of natural and modified CDs in complexing materials. Several reviews and publications provide current information on these complexes and their uses (13-16).

PRODUCTION OF CDS

The enzymatic production of CDs has been recently reviewed (17). The basic process involves standard enzymatic fermentation techniques. All three CDs and some linear oligosaccharides are normally produced. Yields are highly dependent on the source of starch substrate. While most starches can be used, potato starch or an extract of potato starch is often added (18-19). The potato starch component(s) responsible for stimulating CD formation have not been determined. Low starch concentrations (5%-10%) are normally used industrially. Published yields are in the 50% - 80% conversion range.

The distribution of α, β and γ CDs is highly dependent on the origin of biocatalyst used (See section on Cyclodextrin Glucanotransferases). Product distributions may be altered by the addition of specific precipitants, such as aromatics and long chain alcohols (9, 20). These molecules preferentially complex with specific CD species, based on size, and precipitate from solution.

Since a mixture of CDs and linear oligosaccharides result from the enzymatic fermentation with starch, purification is needed to obtain specific CDs. Aromatic organic solvents, such as trichloroethylene or toluene, quantitatively precipitate CDs, allowing simple separation by centrifugation or filtration. The solvents are commonly removed by steam distillation. A non-solvent alternative separation process employs specific alkanophilic *Bacillus sp.* at high pH. Following reaction, the mixture is concentrated and a small amount of CD is added to seed crystallization of the CDs. CD yields from this process are approximately half those of solvent processes.

Other production methods have also been published, such as CGTase immobilization (21, 22), ultrafiltration (23) and the use of isoamylase to increase CD yield. Variations of purification methods include the addition of glucoamylase to degrade non-CD starch hydrolysates to simply separation and the use of various synthetic ion exchange resins in chromatographic separations (24-26) and affinity columns (27).

Despite recent advances in the chemical synthesis of CDs (11), the predominant industrial means of CD production will remain enzymatic. Therefore, improvements of the CGTase biocatalyst to increase the efficiency of CD synthesis and decrease costs is a high priority.

CYCLODEXTRIN GLUCANOTRANSFERASES

Amylolytic glucosylases are probably the most widely studied of the carbohydrate enzymes, due to their biological and industrial importance, as well as their ubiquitous distribution. Many different types of amyloglucosylases exist, with a variety of physical, chemical and catalytic properties. Recent reviews (28-30) describe the classification, characterization, action patterns and biochemistry of different enzymes.

Cyclodextrin glucanotransferase (CGTase) [EC 2.4.1.19] is one of the more unusual members of the amylolytic glucosylase family. Whereas most glucosylases predominantly catalyze a single reaction (either hydrolytic or synthetic), CGTase possesses both capabilities. CGTases produce cyclodextrins with 6, 7 and 8 glucosyl residues and a variety of linear maltooligosaccharides, via disproportionation and coupling reactions (31). While this wealth of catalytic capability provides a unique opportunity to study glycosylase mechanisms, it also is a tremendous industrial problem, incurring high costs due to low CD yields and product separation.

CGTases from several microbiological species have been isolated, however the predominant genus is *Bacillus*. CGTases are extracellular enzymes of approximately 70,000 molecular weight and have been reported as dimers. Most require calcium and have pH optima in the slightly acidic range, similar to α amylases. More recently, several alkanophilic CGTases have been discovered with pH optima of 9-10. Temperature optima range from 50°C to 60°C.

Most CGTases catalyze the formation of all 3 types of CDs, with specific equilibrium distributions. *B. macerans* (32), *B. stearothermophilus* (33) and *Klebsiella pneumoniae* (34) are categorized as α CGTases, producing primarily α CD. Alkanophilic *Bacillus sp.* (35,36), *B. ohbensis* (37), *B. megaterium* (38), *B. circulans* (39) and *Micrococcus sp.* (40) are considered β CGTases. The only CGTases classified as γ CD producers are *Bacillus subtilis* 313 (41) and alkanophilic *Bacillus* 290-3 (20). Through the use of various precipitants, the product distribution of CDs can be altered (20, 42, 43)

CGTases from approximately 13 organisms have been reported. Genes for several CGTases have been isolated and cloned in *E. coli* (See Table II.). The use of genetic expression promoters has been reported to increase cloned CGTase protein expression several thousand-fold (67).

Table II. Cloned CGTases

Organism	Reference
Bacillus 290-3	20
Bacillus macerans	32, 33
B. stearothermophilus	33
Klebseilla pneumoniae	34
Bacillus 38-2	35
Bacillus 1011	36
B. subtilis No. 313	41
Bacillus circulans	44
Bacillus 17-1	45
Bacillus 1-1	46

Two references of X-ray crystallographic data for CGTases have been published (47, 48). Only one of these presents a structural model, for *Bacillus circulans* CGTase(48). This structure has been divided into five subdomains, labeled A-E (See Figure 2). The largest domain is at the N-terminal and forms a $(\beta\backslash\alpha)_8$ TIM barrel structure similar to triose phosphate isomerase (49). Domain B is inserted in this structure following the third β strand. Domain C consists of a greek key β sheet structure. Domains D and E are antiparallel β sheets, with the former having an immunoglobulin topology (50).

Significant primary sequence homology (> 70%) exists among the *Bacillus sp.* CGTases (51). Interestingly, *K. pneumoniae* CGTase only shows about 30% overall homology despite similar molecular size. However, predicted secondary protein structure among CGTases from different origins indicates a high degree of structural homology, despite differences in primary sequence (51).

Interestingly, primary sequence comparisons of CGTases to various α amylases indicate moderate sequence homology, roughly 15% and 25%. Specific regions of high sequence homology have been identified as important catalytic sites and calcium binding sites for the α amylases. However, α amylases primarily only catalyze the hydrolysis of glucans, whereas CGTases also catalyze the cyclization and ligation of oligosaccharides. Despite limited primary sequence homology, predicted secondary structure similarity is very high (51). This similarity presents a unique opportunity to investigate this family of biocatalysts by direct comparison of structural and functional characteristics of different, naturally-occurring enzymes.

SIMILARITIES BETWEEN CGTASE AND A AMYLASE.

Amylases are ubiquitous in Nature and are important for the utilization of carbohydrates in many industrial processes (brewing, baking, sweeteners, alcohol fuels, etc.). α amylases are generally calcium-dependent proteins of approximately 50,000 molecular weight (29). They catalyze the hydrolysis of starch, glycogen and other α-1,4-glucans to form

maltooligosaccharides.   Despite the functional
similarity among α amylases from various
mammalian and bacterial sources, primary sequence
homology is quite low, on the order of about 5%
(52).   It is noteworthy that this homology is
concentrated in 3 short regions for all these
enzymes.   More recently, predicted secondary
structures for several amylases based on primary
sequence data have been found to be similar (53).

Published 3-dimensional structures for porcine
pancreatic α amylase (54) and Taka-α amylase (55)
contain three subdomains, A-C.   The amino terminal
domain (A) is a TIM barrel structure (49) with
antiparallel β sheets inserted between the 3rd β
strand and the 3rd α helix of the barrel (domain
B).   The C domain is a set of 8 antiparallel β
sheets in a greek key topology (56).

The 3 short regions of high primary sequence
homology were all found to be located in either
the catalytic site or the calcium binding site.
Therefore, it has been hypothesized that these
conserved regions are required for the catalytic
actions of these glucosylases.

Structural analyses from X-ray models and
predicted super-secondary models imply that CGTase
structure is primarily a super-set of α amylase
structure.   The active sites and calcium binding
sites of α amylase are believed to reside in the
$(\beta\backslash\alpha)_8$ barrel and β strand loops of domains A and
B.   The various β strand loops of the $(\beta\backslash\alpha)_8$
barrel are also purported to be involved in starch
binding.   The antiparallel β sheet (domain C) is
hypothesized to possess starch binding capability.

To this amylase structure, CGTase adds 2 other
subdomains (D & E), which are antiparallel β
sheets at the C-terminal end of the protein.   The
structure of domain D is similar to immunoglobulin
topology.   The functional role of these domains is
not known.   However, proximity of domain E (an
antiparallel β sheet structure) to the proposed
catalytic site, suggests a role in substrate
binding and conformation.

COMPARISON OF AMYLOLYTIC GLUCOSYLASES

While the literature is rich in scientific
information on glucosylases, recent interest has
focused on the hypothesis that all these enzymes
share a common catalytic mechanism, despite
differences in their product specificity (57).
Indeed, it has been proposed that all glycosylases
share the same basic chemical mechanism (58).   The
α amylases have been the focus of much of this
attention, as the primary protein sequence (59),
tertiary protein structure (54,55) and catalytic
mechanism (57) have been recently delineated.

Recently, attention has focused on analyzing the structural and functional relationships of these biocatalysts. Despite limited primary sequence homology among the various enzymes (60-62), predicted secondary structure (51) and limited X-ray crystallographic structures (48, 54, 55) indicate that all these proteins are basically variations of the same structure (($\beta\backslash\alpha)_8$ TIM barrel, See Figure 3). This suggests similar mechanisms, albeit with differences in product specificity and substrate binding. An X-ray model of xylose isomerase (63) and predicted secondary structure of α-glucosidase (51) also indicate the same ($\beta\backslash\alpha)_8$ barrel structure. It has been speculated that a similar extension might be made to other glucosylases such as amylosucrase and dextransucrase (64).

GENETIC MUTATIONS OF AMYLOGLUCOSYLASES

Based on primary sequence and X-ray crystallographic structures of amylases and CGTases, it is believed that the overall tertiary structure of amyloglucosylases is conserved. The basic α amylase TIM barrel construction with β strand loops between the helices and β sheets of the ($\beta\backslash\alpha)_8$ barrel structure and large antiparallel β sheet domain appear to be universal elements of these proteins. The conservation of primary sequence regions involved in catalysis and calcium binding throughout these proteins implies a common mechanism. However, distinct variations in β strand loop structures are believed to be responsible for differences in product specificity of the different α amylases. The additional domains present in CGTases, along with variations in the β strand loop structures, are hypothesized to be responsible for the cyclization phenomena of CGTases and the ability to couple glucosyl residues.

Current research is exploring the effects of altering these active sites and binding structures to "engineer" new catalytic specificity and modify biocatalyst behavior. Site-directed mutagenesis on alpha amylase (65) have been used to identify catalytic amino acid residues. Similar work on glucoamylase (66) has been performed, with current work focusing on changing the catalytic action of this enzyme. Schmid (67) noted that deletion of part of the E subdomain of *K. pneumoniae* CGTase reduced the activity of its CGTase, but did not alter its product specificity. In our laboratory we have recently cloned CGTase genes from several *Bacillus sp.* and expressed native CGTase in *E. coli* and yeast. Using PCR (polymerase chain reaction) recombinant genetic mutation methods, we are creating site-directed mutants of the β strand loops in domain B of these CGTases to investigate the role of specific residues in the catalytic properties of this enzyme. We are also creating regional deletion of subdomains D and E to determine if the resulting protein retains conventional hydrolytic capability. It is our hope that we will eventually be able to unravel the synthetic and hydrolytic structural-functional relationships of CGTases and begin to apply these principles to build more effective glucan biocatalysts.

CONCLUSIONS

Studies of observed and predicted structure point to the conclusion that despite significant differences (> 70%) in primary sequence, α glucosylases form a family of biocatalysts with very similar protein tertiary and secondary structure. Small variations in specific areas of protein structure account for very different product specificity, due to changes in substrate binding. The presence of additional domains, in conjunction with some variation in sequence homology, can produce dramatic product configurational changes, as witnessed by CGTase-catalyzed cyclization and multiple product specificity.

By direct comparison of structural and functional characteristics of different, naturally-occurring enzymes and modified biocatalysts, researchers hope to delineate the structural features controlling the hydrolytic and synthetic catalytic activities of this family of glucosylases. This understanding will eventually lead to the ability to "engineer" glucan biocatalyst capability.

These engineered biocatalysts will in turn provide new means to efficiently utilize natural biopolymers to synthesize chemical intermediates, fuels, food products, pharmaceuticals and plastics. New synthetic biocatalysts will allow the creation of a host of new biopolymers possessing new functions and properties for industrial, pharmaceutical and agricultural uses.

ACKNOWLEDGMENTS

This work has been supported by the National Science Foundation No. 8908391-BCS and Purdue University.

LITERATURE CITED

1. Villiers, A. Compt. Rend. Acad. Sci. 1891, 112, 536.
2. French, D. Adv. Carbohydr. Chem. 1957, 12, 189.
3. Szejtli, J. In Cyclodextrin Technology; Davies, J. E., Ed.; Kluwer Academic Publishers: Dordrecht, 1988; Chapter 3-6.
4. Cramer, F. In First Proc. Intl. Symp. Cyclodextrins; Szejtli, J., Ed.; Reidel: Dordrecht, 1982; p 3.
5. Armstrong, D. W. Anal. Chem. 1987, 59, 84A.
6. Cramer, F. In Cyclodextrins and their Industrial Uses; Duchene, D., Ed.; Editions de Sante: Paris, 1987; p 11.
7. Szejtli, J. Tibtech 1989; 7, 170.

8. Bender, M. L. and Komiyama, M. In Cyclodextrin Chemistry; Springer-Verlag: New York, 1978; p 2.
9. Szejtli, J. In Cyclodextrin Technology; Davies, J. E., Ed.; Kluwer Academic Publishers: Dordrecht, 1988; Chapter 6.
10. Krechl, J. Castulik, P. Chemical Scripta; 1989, 29, 173.
11. Takahashi, Y., Ogawa, T. Carbohydr. Res.; 1987, 164, 277.
12. Mori, M. Ito, Y., Ogawa, T. Tetrahedron Lett.; 1989; 30, 1273.
13. Vaution, C., Hutin, M., Glomot, F. and DuChene, D. In Cyclodextrins and Their Industrial Uses; DuChene, D., Ed.; Editions de Sante: Paris, 1987; Chapter 8.
14. Special Edition on Cyclodextrins, Carbohydr. Res. 1989, 192.
15. Cyclodextrin News; Szejtli, J. and Pagington, J., Eds.; FDS Publications: P.O. Box 41 Trowbridge, Wiltshire England.
16. Proc. of the 5th Intl. Symposia on Cyclodextrins; March 28-30, 1990, Paris, to be published by Kluwer Academic Publishers: Dordrecht.
17. Sicard, P. J., Saniez, M.-H. In Cyclodextrins and Their Industrial Uses; DuChene, D., Ed.; Editions de Sante: Paris, 1987; Chapter 2.
18. Matsutani Kagaku Kogyo Japanese Patent 9224 1971.
19. Nihon Shokuhin Kako; Rikagaku Kenkyusho Japanese Patent 92288 1974.
20. Schmid, G. Huber, O. Eberle, H.-J. In Proc. 4th Intl. Symp. on Cyclodextrins; Huber, O., Szejtli, J., Eds.; Kluwer Academic Publishers: Dordrecht, 1988; p 87.
21. Kato, T. and Horikoshi, K. Biotech. Bioeng. 1984, 26, 595.
22. Crump, S. P., Rozzell, J. D. In Proc. 4th Intl. Symp. on Cyclodextrins; Huber, O., Szejtli, J., Eds.; Kluwer Academic Publishers: Dordrecht, 1988; p 47.
23. Hokse, H. U.S. Patent 4477568 1984.
24. Hokse, H. J. Chromatogr. 1980, 189, 98.
25. Rikagaku Kenkyusho, Nihon Shokuhin Kako European Patent Spec. 45464 1985.
26. Nihon Shokuhin Kako, Rikagaku Kenkyusho U. S. Patent 4418144 1983.
27. Mattsson, P., Makela, M., Korpela, T. In Proc. 4th Intl. Symp. on Cyclodextrins; Huber, O., Szejtli, J., Eds.; Kluwer Academic Publishers: Dordrecht, 1988; p 65.
28. Vihinen, M.; Mantsala, P. In Critical Reviews in Biochemistry and Molecular Biology; 1989, 24, 329.
29. Robyt, J. F. In Starch: Chemistry and Technology, 2nd edition; Whistler, R. L.; BeMiller, J.; Paschall, E. F., Eds.; Academic Press: New York, 1984; p. 87.
30. Cote, G.; Tao, B. Glycoconjugate J.; in press.
31. French, D.; Levine, M.; Norberg, E.; Nordin, P.; Pazur, J.; Wild, G. J. Am. Chem. Soc. 1954, 76, 2387.
32. Takano, T.; Fukuda, M.; Moma, M.; Kobayashi, S.; Kainuma, K.; Yamane, K. J. Bacteriol. 1986, 166, 1118.

33. Sugimoto, T.; Kubota, M.; Sakai, S. UK Patent GB 2169902 1986.
34. Binder, F.; Huber, O.; Bock, A. Gene 1986, 47, 269.
35. Kaneko, T.; Hamamoto, T.; Horikoshi, K. J. Gen. Microbiol. 1988, 134, 97.
36. Kimura, K.; Takano, T.; Yamane, K. Appl. Microbiol. Biotechnol. 1987, 26, 149.
37. Yagi, Y.; Iguchi, H. Japan Patent 74124385 1974.
38. Okada, S.; Tsuyama, N. U. S. Patent 3812011 1974.
39. Nakamura, N.; Horikoshi, K. Agric. Biol. Chem. 1976, 40, 1785.
40. Yagi, Y.; Kuonon, K.; Inui, T. Eur. Patent 0017242 1983.
41. Kato, T.; Horikoshi, K. Agric. Biol. Chem. 1986, 50, 2161.
42. Armbruster, F. C. In Proc. 4th Intl. Symp. on Cyclodextrins; Huber, O., Szejtli, J., Eds.; Kluwer Academic Publishers: Dordrecht, 1988; p 33.
43. Seres, G., Barcza, L. In Proc. 4th Intl. Symp. on Cyclodextrins; Huber, O., Szejtli, J., Eds.; Kluwer Academic Publishers: Dordrecht, 1988; p 81.
44. Nitschke, L. Diplomarbeit, Universitat Freiburg i.Br 1989.
45. Horikoshi, K. In Proc. 4th Intl. Symp. on Cyclodextrins; Huber, O., Szejtli, J., Eds.; Kluwer Academic Publishers: Dordrecht, 1988; p 87.
46. Schmid, G., Englbrecht, A., Schmid, D. In Proc. 4th Intl. Symp. on Cyclodextrins; Huber, O., Szejtli, J., Eds.; Kluwer Academic Publishers: Dordrecht, 1988; p 71.
47. Kubota, M.; Mikami, B.; Tsujisaka, Y.; Morita, Y. J. Biochem. 1988, 104, 12.
48. Hofmann, B.; Bender, H.; Schulz, G. J. Mol. Biol. 1989, 209, 793.
49. Banner, D. W.; Bloomer, A. C.; Petsko, G. A.; Phillips, D. C.; Pogson, C. I.; Wilson, I. A.; Corron, P. H.; Furth, A. J.; Milman, J. D.; Offord, R. E.; Priddle, J. D.; Waley, S. G. Nature (London) 1974, 255, 609.
50. Schiffer, M.; Girling, R. L.; Ely, K. R.; Edmundson, A. B. Biochemistry 1973, 12, 4620.
51. MacGregor, A.; Svensson, B. Biochem. J. 1989, 259, 145.
52. Nakajima, R.; Imanaka, T.; Aiba, S. Appl. Microbiol. Biotechnol. 1986, 23, 355.
53. MacGregor, E. A. Protein Chem. 1989, 7, 399.
54. Buisson, G.; Duee, E.; Haser, R.; Payan, F. EMBO J. 1987, 6, 3908.
55. Matsuura, Y.; Kusunoki, M.; Harada, W.; Kakudo, M. J. Biochem. 1984, 95, 697.
56. Richardson, J. S. Adv. Protein Chem. 1981, 34, 167.
57. Tao, B.; Reilly, P.; Robyt, J. Biochem. Biophys. Acta 1989, 995, 214.
58. Hehre, E.; Okada, G. Genghof, D. Adv. Chem. Ser. 1973, 117, 309.
59. Pasero, L.; Mazzei-Pierron, Y.; Abadie, B.; Chicheportiche, Y.; Marchis-Mouren, G. Biochim. Biophys. Acta 1986, 869, 147.

60. Sakai, S.; Kubota, M.; Yamamoto, K.; Nakada, T.; Torigooe, K.; Ando, O.; Sugimoto, T.; Denpun Kagaku 1987, 34, 140.
61. Svensson, B. FEBS Let. 1988, 230, 72.
62. Svensson, B.; Jespersen, H.; Sierks, M.; MacGregor, A. Biochem. J. 1989, 264, 309.
63. Carrel, H. L.; Rubin, B. H.; Hurley, T. J.; Glusker, J. P. J. Biol. Chem. 1984, 259, 3220.
64. Tao, B. Y.; Reilly, P. J.; Robyt, J. F. Carbohydr. Res. 1988, 181, 163.
65. Vihinen, M.; Ollikka, P.; Niskanen, J.; Meyer, P.; Suominen, I.; Kearp, M.; Holm, L.; Knowled, J.; Mantsala, P. submitted.
66. Sierks, M.; Svensson, B.; Ford, C.; Reilly, P. Personal communication
67. Schmid, G. Tibtech 1989, 7, 244.
68. Richardson, J. In The Protein Folding Problem; Wetlaufer, D., Ed.; Am. Assn. Adv. Sci. Selected Symp. No. 89: Washington, DC, 1984; Chapter 1.

Figure 1. Structure of Cyclodextrins

Figure 2. X-ray Crystallographic Structure of *B. circulans* CGTase

Figure 3. Model of $(\beta\backslash\alpha)_8$ TIM barrel from triose phosphate isomerase

OPTIMIZATION OF PLANT TISSUE CULTURE SYSTEMS

Subhas C. Mohapatra, Dae-Weon Lee and William F. McClure

Laboratories of Biotechnology and Bioinstrumentation
Department of Biological and Agricultural Engineering
North Carolina State University, Raleigh, NC 27695-7625

Optimization of any commercial process entails maximum unit of work done per unit time at minimum cost. Because time and cost are the limiting factors for the optimization of any system, automation is a necessary component of optimization. This is true for the optimization and commercialization of plant tissue culture, which has become an integral part of molecular biology and genetic engineering because of its usefulness for gene transfer and transgenic plant production. It is also a powerful tool for biotechnology and as such carries considerable potential for commercial propagation of tissues and production of. In order for this potential to be fully exploited, plant tissue culture must be elevated from a laboratory technique to an industrial process through automation. The above laboratories have recently been involved in the application of tissue culture for mass production of seedlings. Seedling production through tissue culture is usually accomplished through solid-agar culture and consists of the following general steps: a) Sterilization and handling of chemicals, glassware and tissues/organs of interest, b) Growth and differentiation of the tissue under culture and c) handling and growth of the resultant seedlings. In order for tissue culture to be optimized for commercial application, each of the above steps must be automated to the extent possible. It appears that many of the automation already developed for chemical and glassware handling for animal and microbial cell culture can be adapted to plant tissue culture without much difficulty. This leaves automation of tissue and seedling handling and growth conditions to be optimized. Both of these

111

aspects are under investigation in our laboratories and will be addressed in this paper.

SEEDLING PRODUCTION

Seedling production through tissue culture can be accomplished through either direct differentiation of the tissue explant or differentiation of the callus produced from the explant. Light is indispensable for both methods. Thus, tiers of shelves each with a bank of over-head lights is a common feature in most tissue culture laboratories. This, however, creates high energy demand, especially during the peak hours. A primary objective of our optimization efforts is reduction of energy demand through reduced light requirement. Potential for this was explored by investigating the effect of the light period on tissue differentiation. Results from this study showed that one hour of light (200 μmol.m^{-2}.sec^{-1}) during a 24-hour period was sufficient to induce seedling formation in tobacco leaf explants. It was further observed that the one-hour light requirement can be supplied through several intermittent exposures rather than one continuous exposure. This permitted mounting of the tissue culture flasks on a rotary station under an over-head light (350 μmol.m^{-2}.sec^{-1}) such that each flask was exposed to light for 10 sec every three min. This approach permitted not only to reduce energy demand thorough elimination of multiple banks of light, but at the same time permitted use of vertical space without having to use multiple shelves. Rotating the tissue cultures around its vertical axis resulted in different orientations of the flasks with respect to the earth's gravity. This effect was, presumably, canceled biologically because of the reversed orientation of the tissues at two opposite points on the rotary tissue culture station. This mutually canceling effect is designated as gravity compensation, and growth and differentiation of tissue explants under gravity compensation is under further investigation.

TISSUE AND SEEDLING HANDLING

Following growth and differentiation of the tissue explant under culture, the seedlings have to be harvested, singulated and transferred to trays for subsequent growth until shipment and planting, and each of these steps must be automated to the extent possible. While several options are available for automating tissue and seedling handling, our objective is to accomplish the task through "intelligent robotics" incorporating computer vision, image analysis, identification, gripping/releasing, and targeted movement. Since these operations for tissue culture would involve complex algorithms and control mechanisms, some of the concepts and basic principles were tested through application to a simpler system, yet with the same general objective. This test-system consisted of seedlings produced through seed germination rather than tissue culture. Rarely does a seed lot, no matter how strong is its vigor and viability, give 100% germination,even under relatively more favorable conditions inside a greenhouse. Thus, a certain percentage of cells in a given seedling tray will remain empty and need to be filled with seedlings from another tray.

Work has already shown that imaging spectroscopy can differentiate plant tissues from the surrounding growth media. Furthermore, plant tissue details are also refined by imaging spectroscopy to the extent that this technique appears to have the potential for differentiating cluster excisions. It is thus conceivable that the aforementioned robot would be able to recognize the donor and receptor trays, and their cells with or without seedlings. Further, the artificial intelligence of the robot could perhaps be programmed to distinguish seedlings of different growth potential. Tobacco seedling was selected as the initial experimental material because of its common use in tissue culture research and because it is a major cash crop in the Carolinas and Virginia.

Video images were captured with the aid of a Panasonic WV-CD50 camera and displayed a Panasonic MW-5410 monitor. Video data were digitized with a Panasonic DT2851 Translator and stored in a computer (IBMPC/AT). Imaging and digitization of the image was performed on the seedling, and the tray-cell with and without seedling. Imaging spectroscopy was achieved by interposing narrow band interference filters between the camera and the trays such that the camera viewed the trays through the selected filters. The filters were mounted on a stepper motor mechanism so that the filters could be easily changed manually or by the computer. Image digitization was accomplished by counting the number of pixels in the image of interest.

The digital information was subsequently used to guide the Cartesian Coordinate robot (hereafter designated as Agricultural Robot or AGBOT) to the location occupied by the tray and individual cells in the tray with or without seedlings. A stepper motor control card (PCMC, Advanced Micro System) installed in the computer was used to control the movement of the AGBOT. The final step in the automation of seedling manipulation is the development of a gripper to pick up and release the seedling plug. The actuator of the gripper uses a stepper motor to close and open the four fingers. Two types of fingers were tested to grip and release the seedling plug within one square inch. One type of finger uses four spades and another type uses four needles to grip the plug. The spade finger is made of spring steel and shaped to fit the profile of the tray cells. The spades work well where the growth media is loose and must be retained along with the seedling being transferred. The needle fingers were designed to pick up compact composition-type media which does not fall apart when lifted.

ALTERNATIVE PLANT TISSUE CULTURE SYSTEMS

Roy E. Young and S. Andy Hale

Agricultural Engineering Department

Clemson University, Clemson, SC 29634-0357

INTRODUCTION

Growth medium for plant tissue culture may assume liquid (suspension) and solid (agar) forms, yet agar is most widely used both in laboratories and in commercial settings. In fact, except for scale, commercial tissue culture facilities strongly resemble research laboratories in equipment and procedures. Propagation containers are numerous, individual and small. They consist of test tubes, Petri dishes, flasks and jars. Large numbers of small containers make labor intense, routine, tedious, and often dull, particularly on the commercial scale (Wochok, 1979).

Both agar and liquid suspension cultural methods are characterized by continual depletion of medium. Furthermore, frequent, highly labor-intensive transfers are required every 4-8 weeks. Harvested product cost ranges from $0.12 to $0.17 per micropropagule. Labor accounts for 40-90% of this total operating cost. Competitiveness of tissue culture products is currently limited to crops with high profit margins such as ornamentals or other plants with superior properties. Tissue culture propagation faces a substantial challenge to be competitive compared to conventional propagation of bedding plants (< $0.05 US per cutting) or of agronomic and forestry crops (< $0.01 US per cutting). Sluis and Walker (1985) projected that

115

bridging this cost gap for growth in the industry will depend heavily upon advances in mechanization and automation.

REVIEW OF LITERATURE

Maene and Debergh (1985) tried to reduce manual labor by adding liquid media to established, exhausted agar cultures instead of transferring plant material to new vessels with fresh agar media. Through media modifications they were able to achieve elongation and enhanced rooting, but they encountered increased vitrification from submersion in the liquid. Aitken-Christie and Jones (1987) applied twice weekly flushings of liquid medium over agar for periods of 4-6 hours to produce radiata pine shoot hedges *in vitro*. Shoots retained on the initial agar media and flushed with liquid media had shoot elongation significantly greater than shoots conventionally transferred to fresh agar media. Aitken-Christie and Davies (1988) expanded their earlier jar cultures of shoot hedges to a large (390 mm l x 250 mm w x 120 mm h), clear, autoclavable, polycarbonate container with automatic addition and removal of liquid nutrients controlled by peristaltic pumps and programmable time clocks. Problems with sterility were encountered under normal laboratory laminar flow bench conditions.

Tisserat and Vandercook (1985,1986) constructed an automated plant culture system (APCS) which consisted of a 2-piece polystyrene chamber to which periodic floodings of liquid nutrient were controlled by a computer. Explants of orchids, carrot, date palm, aster and cow tree, positioned within compartments of plastic ice cube trays with holes drilled in their bottoms, were immersed by nutrient flooding every 2 hours. Nutrient was then allowed to gravity flow back into the supply reservoir, leaving the plant tissue to rest on glass beads and aerate (Wood, 1985). Growth of plants either equaled or exceeded that of plants on concurrent agar media cultures. After 9 months of culture in the APCS, orchid tips produced 4 times as much tissue as orchids transferred every 6 weeks to fresh agar media. Simonton and Robacker (1988) developed a liquid nutrient system which supplied several 10-liter polycarbonate containers as separate culture vessels from a central supply carboy. They had limited success with a chemical sterilization scheme of detergent, alcohol and sterile water rinses between culture cycles to avert time-consuming and difficult autoclavings. Farrell (1987) reported that Agro-Clonics, Inc., Orange, CA. had an automated plant micropropagation system by the trademark of Mega-Yield. This system used a synthetic hydrophilic hollow fiber material adapted from the bio-medical industry to mimic the plant's xylem fibers to wick liquid nutrient media.

Levin (1983, 1984) and Levin, et. al. (1988) described the process and components for an automated plant tissue culture system marketed by the Isracli company P.B.Ind. under the identity of Vitromatic. The system was liquid nutrient based. The process used a homogenizer device (similar to the concept offered by Cooke (1979) for ferns) to separate densely growing meristematic tissue, a sieving apparatus to obtain small clusters of tissue of fairly uniform sizes and free of debris, and a bulk tank to dilute sized tissue in a sterile aqueous medium for dispensing into individual culture vessels for plantlet development. The separation (homogenizer) step was eliminated with embryogenic cultures.

A nutrient mist bioreactor for plant tissue culture was developed by Weathers and Giles (1988). Tissue was grown on a biologically inert, fine mesh screen within a sterile chamber. A nutrient mist was sprayed from above onto the propagating tissue. Fox (1988) described a commercial version called the Mistifier being developed by BioRational Technologies (Stow, MA) and to be manufactured and marketed by Manostat (New York, NY). Growth rates of 3.5 times greater in the mist bioreactor than on conventional agar cultures were cited. It was also observed that callus in the mist bioreactor exhibited a bulbous, almost spherical growth pattern compared to rather erratic growth patterns on agar.

Hamilton, et. al. (1985) reported a technique for growing micropropagules on flat microporous polypropylene membranes floating on liquid nutrients. Sinking of the membrane as the weight of the plant material exceeded its buoyancy was observed as a problem. Kong and Chin (1988) observed improved growth of asparagus protoplasts on microporous polypropylene membrane. They surmised that the technique allowed the cultures to receive both aeration and nutrients without agitation needed with suspension cultures. Matsumoto and Yamaguchi (1989) used several synthetic nonwoven materials as supporting agents for *in vitro* culture of banana protocorm-like bodies.

For reuse and plant growth, a 50% acetate and 50% polyester fiber gave best compromise performance, considerably superior to agar. Ford (1989) described a flexible, gas-permeable, transparent plastic film package commercially available for plant tissue culture under the trademark StarPac.

METHODS AND EQUIPMENT

Continual-Flow, Liquid Nutrient Vessels

Unlike liquid suspension cultures where plant tissue is immersed in static or oscillating liquid in a sealed vessel, the concept of continual-flow flushes the vessel with fresh liquid nutrient periodically. The plant tissue does not have to be disturbed by transfer to another vessel to receive fresh nutrients. The fresh nutrient is brought to the undisturbed plant tissue.

We developed an apparatus using modified, 7-inch diameter, heavy duty glass beakers and a peristaltic pump to transfer media from a supply carboy to the beakers. The beakers were closed by flat glass tops seated on flat ground beaker rims. Liquid media entered each beaker through a port in the sidewall of the beaker approximately 50 mm above the bottom. This port was sufficiently high above the bottom of the beaker to assure a physical discontinuity of liquid between the beaker and the supply lines to restrict contaminant transport. Spent media was drained at desired intervals by gravity flow through the bottoms of the beakers.

Alternative Support Materials

The most apparent deficiency of liquid suspension culture compared to agar culture is its inability to support the tissue so that it receives sufficient aeration for non-vitrified growth. Consequently, as we initiated our work with continual-flow liquid culturing, we surveyed various materials as alternatives to agar for tissue support. Based on criteria such as autoclavability, strength, absorbancy, inertness to plant tissue and reusability, we screened approximately 30 material options ranging from paper to plastics to wire screening. Five materials were selected for testing: polyurethane foam, non-woven polypropylene fiber, glass beads, microporous polypropylene membrane (Celgard^R membrane by Hoechst Celanese, Separation Products Division, Charlotte, NC) and polypropylene netting.

Each of the materials were tested for growth support in both static vessels and in continual-flow vessels. The static liquid medium tests were conducted in Magenta GA-7 containers, square plastic vessels designed specifically for the plant tissue culture industry.

The polypropylene netting was tested both as an alternative support to agar and as a device through which to train growth of plantlets. We have used primarily an 8 x 8 mesh (i.e., 8 grids per inch in a square pattern). A trifold configuration of this netting into a square approximately 50 x 50 mm was initially used with explants placed between the bottom and middle layers. As the explants callused and produced shoots, they grew through the top two layers. We examined separating the plantlets from the callus and regenerating tissue by two actions: (1) peeling the top layer away from the middle layer and (2) shearing the top layer laterally past the middle layer. The trifold netting was used in the continual-flow, liquid nutrient vessels and Magenta vessels both on top of other support materials and alone.

Later, we worked with devices that combined the two materials, microporous polypropylene membrane and netting. One boat-like device consisted of a rigid, open polypropylene frame with a microporous membrane bottom. A second open frame was wedged inside the sidewalls of the boat to tighten netting 5-10 mm above the membrane. In a second "sandwich" device, Figure 1, polypropylene netting was attached to the top edge of a solid sidewall frame and the membrane was attached to the bottom edge. In both devices, the explant was dropped through the grid of the netting for culture initiation. Plantlets grew up through the netting where they were accessible for shoot cutting and singulation.

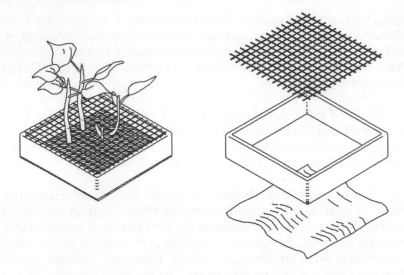

Figure 1. Sandwich type membrane-netting device.

Most recently we have been interested in exploring use of the gas-permeable, transparent polyethylene film manufactured by AgriStar under the tradename StarPac. To test its potential vapor permeability, we tested five thicknesses (0.4, 0.6, 0.8, 1.0 and 1.25 mil) in four configurations: (1) StarPac containing agar medium, (2) StarPac containing liquid medium, (3) StarPac in external liquid medium in Magenta GA-7 vessels and (4) StarPac in external liquid medium in the continual-flow system. Tobacco (*Nicotiana tobacum*) leaf discs were used as explants to initiate cultures in the sealed pockets of the StarPacs. If this transparent, gas-permeable film were also vapor-permeable, sealed, aseptic packages could be made for the continual-flow, liquid culture system to reduce contamination and to minimize vitrification.

Plant Materials and Media Preparation. Tomato (*Lycopersicon esculentum* cv. Rutgers) seeds were disinfected by soaking in 10% Clorox for 10 minutes and rinsed twice in sterile, distilled water. Seeds, 15-20 per plate, were then placed in sterile glass Petri plates containing filter paper and water. The plates were incubated at 27°C with a 16 hour daylength. Cotyledons were aseptically removed after 7-10 days germination and planted in vitro in Magenta GA-7 vessels and the continual-flow beaker system. Fresh weights were recorded as grams per vessel. Culture media used was Murashige and Skoog salts (Murashige and Skoog, 1962) with 30 g/l sucrose and 2 mg/l zeatin. Growth data were recorded after 12 to 21 days as fresh weight gain. No dry weights were recorded.

Subcultures of begonia (*Begonia rex*) were made with small clusters (0.2 to 0.7 g. fresh weight) on Murashige and Skoog salts containing 30 g/l sucrose, 0.17 g/l NaH_2PO_4, 0.08 g adenine sulfate, 0.1 g/l i-innositol, 0.4 mg/l thiamine HCl, 10 M indole acetic acid, 10 M kinetin at a pH of 5.7. Seven g/l of agar was used in agar treatments with 40 ml of that medium added to each Magenta GA-7 vessel. Forty ml of liquid (agarless) medium was used with Celgard[R] microporous membrane rafts (Sigma Chemical Co., St. Louis, MO). Thirty ml of liquid medium was used in Magenta GA-7 vessels containing 5x5x1 cm squares of nonwoven polypropylene fiber treated with Decerisol (a nontoxic surfactant). Thirty ml of liquid medium was also used with 3 mm diameter glass beads at a height of 3 cm in the Magenta vessel. Growth data was recorded after 25-33 days as final fresh weight.

Tobacco (*Nicotiana tobacum*) leaf discs were surface sterilized for 10 minutes in 10% Clorox solution and rinsed twice in sterile, distilled water. Fresh weights were recorded for each disc before it was placed in a pocket of the StarPac which was subsequently heat sealed. The culture media used Murashige and Skoog salts (Murashige and Skoog, 1962) with benzyladenine and 30 g/l sucrose. Cultures were grown at 27°C and a daylength of 16 hours. Fresh weights were measured again after 30 days in culture and fresh weight gains were calculated for each disc.

All the Magenta GA-7 vessels were autoclaved with the solid matrices and medium in place. For the continual-flow (beaker) vessels, 3 to 4 liters of media were autoclaved in 9-liter carboys for one hour at 121°C and 18 psi pressure. Sucrose and zeatin were filter sterilized in a 150 ml volume through a 500 ml, 0.2 um SuperVac vacuum filtration unit (Gelman 3010). This solution was then quickly added to the supply carboy under the laminar flow hood and mixed. The overriding principle of this procedure was to transfer minimum volumes of medium for a minimum length of time with the smallest vessel openings possible.

It was best to complete the medium sterilization two days prior to culture initiation to monitor for contamination. All remaining system components were steam sterilized individually wrapped in brown, lint-free paper. The paper was removed only under the laminar flow hood when ready to use.

RESULTS AND DISCUSSION

Continual-Flow, Liquid Nutrient System

Because of earlier problems with maintaining aseptic conditions in the continual-flow beaker vessels, we were unable to accumulate sufficient replications to compare statistically significant differences among support matrices by this technique. Therefore, all support matrices comparisons were made from tests conducted with static liquid medium in Magenta GA-7 vessels. We did, however, have individually successful, sterile cultures for all matrices in the continual-flow beakers. Protocols for repeatedly achieving aseptic cultures in the continual-flow vessels have now been established, however.

Support Matrices

Early in our growth performance testing of polyurethane foam, it became evident that the wicking action was too slow. Tomato cotyledons would dessicate before liquid nutrient became available to them. We eliminated polyurethane foam as a viable material. Nonwoven polypropylene fiber wicked liquid more readily than polyurethane foam, but not without non-uniformity of wetting at the surface. Explants frequently dessicated on this material also. A non-toxic surfactant (Decerisol) increased rate of wicking and improved uniformity of wetting for the nonwoven fiber. Mixing and maintaining sterility of the surfactant, however, are additional complications to using the nonwoven fiber. We have also observed examples of regenerative tissue becoming intertwined in the nonwoven fiber, a condition that discourages successful use in a mechanized separation operation. Glass beads provide adequate support of micropropagules, but are very sensitive to maintaining a liquid level sufficiently close to the plant material to avert non-optimal growth or dessication. A precise level of liquid medium, however, is not readily achieved and maintained. Perhaps the greatest detriment to the use of glass beads in a mechanical system is their inability to maintain shape. They must be restrained by a carrier apparatus to which they can comform in shape.

The trifold polypropylene netting functioned similar to glass beads, although it did provide a defined form within itself. Liquid nutrient had to be recycled with intervening periods of draining to avoid vitrification. When used on top of the other alternative support materials tested, trifold netting held the plant tissue too far from the upper surface of the wetting front and resulted in tissue dessication also. Other difficulties encountered with the trifold netting were achieving a flat conformation of the three folds and inserting explants and preventing their falling through the open edges of the device.

The microporous polypropylene membrane provided physical separation of the tissue from the liquid nutrient yet allowed transfer of nutrient to the plant tissue through the micropores. Its buoyancy enabled it to float a small biomass, even as a flat sheet lying on the liquid surface. It also provided essentially 100% surface contact with the liquid media, a feature which improves uniformity of tissue. However, additional buoyancy and/or edge walls are needed to prevent flooding as the tissue mass increases.

Figure 2 compares mean fresh weight gains for initiating cultures of tomato cotyledons and transfers of cultures of begonia in static medium in Magenta GA-7 vessels on four support materials: agar, Celgard[R] microporous membrane, glass beads and nonwoven polypropylene fiber. Four to five replications were made of each material, or matrix, in Magenta vessels containing 4-6 cotyledons of equal initial age whose combined fresh weights were recorded. Cotyledons on microporous membranes had the highest mean fresh weight gain, 0.77 g, significantly greater than tissue grown on traditional agar (0.47 g). In fact, growth on microporous membrane was also significantly greater than growth on glass beads (0.26 g) and nonwoven polypropylene fiber (0.37 g). Fresh weight gain on glass beads (0.26 g) was significantly less than gain on agar. Low growth performance on glass beads was probably a result of using a static liquid level in the Magenta. We observed detrimental effects of both too high or too low

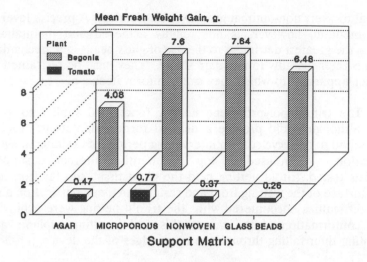

Figure 2. Comparisons of mean tissue fresh weight gains on four alternative support matrices.

liquid medium levels. Vitrification and rapid tissue demise resulted from too high medium and dessication resulted from too low media. Medium level was extremely critical on glass beads. Although not quite as critical, tomato cotyledons were similarly sensitive to medium level on nonwoven polypropylene fiber.

Figure 2 also shows mean fresh weight gains measured for transferred cultures of clusters of begonia tissue. Each of the four support matrices were used in Magenta vessels with static medium, three replications of each support matrix. Five to six clusters of begonia tissue divided from established *in vitro* agar cultures were transferred to each Magenta vessel. Combined weights of clusters in each vessel were recorded. Mean fresh weight gains of 7.60 g on microporous membrane and 7.64 g on nonwoven polypropylene fiber were essentially equivalent and the highest for the four support matrices. A mean fresh weight gain of 4.08 g on agar was the lowest of the matrices tested. Glass beads yielded a weight gain of 6.46 g, which was not significantly different from microporous membrane and nonwoven fiber. All three alternative matrices - microporous membrane, nonwoven fiber and glass beads - yielded significantly higher mean fresh weight gains of transferred begonia tissue than agar. Improved growth performance of begonia on glass beads and nonwoven polypropylene fiber probably resulted from knowledge gained with the earlier tomato cotyledon studies concerning critical levels of liquid medium with these two support matrices. Moreover, the begonia tissue transferred from established in vitro (agar) cultures appeared to be less susceptible to contamination and perhaps more resilient to environmental shock than the initiation cultures of tomato cotyledons.

Other plants successfully cultured on microporous membrane by our horticulturists included Boston fern (2 varieties), staghorn fern, orchids (12 varieties), watermelon, cauliflower, sweet potato, begonia, hostas and cherry. These all showed equal or superior growth to similar tissue on agar. Seven watermelon zygotes on microporous membrane in a Magenta vessel attained fresh weights 6 times greater than 7 zygotes on agar in a similar vessel after 18 days in culture. After 15 weeks culture, 8 Magenta vessels with 5 orchid protocorms each on microporous membrane had 10 times as much fresh weight per vessel as 2 Magenta vessels of agar with 5 protocorms each. Liquid media formulations have been the same as that used in the agar-gelled media except without the agar.

The StarPac film tests clearly indicated that none of the configurations or thicknesses of the polyethylene film were vapor permeable. No appreciable fresh weight gains were observed in configurations with the medium external to the film for any thickness of film. Fresh weight gains were evident for both agar and liquid medium inside the StarPacs.

Initial weight gains were visibly greater for discs in liquid inside the StarPac than for discs on agar inside the StarPacs. However, as the culture period progressed, this trend reversed. Final fresh weight gains were significantly greater on agar than in liquid. Apparently, vitrification and subsequent tissue deterioration accounted for this reversal.

Mechanical Separation

The trifold netting enabled us to examine two ways of mechanically separating the plantlets from the calli and poorly differentiated tissue. When the upper layer of netting was lifted vertically, or peeled away from the middle layer, plantlets did not separate cleanly from the calli, but tended to strip through the grids of the netting and become damaged. On the other hand, separation of the plantlet from the calli was easily achieved by shearing the upper netting laterally past the middle layer.

The membrane-netting device shown in Figure 1 has enabled separation of plantlets from regenerating tissue by cutting along the upper surface of the netting. The cut plantlets may subsequently be treated as unrooted cuttings and potentially transferred to nursery media after receiving rooting hormone treatment. We envision that the "sandwich" device might become sufficiently cost effective to be disposable after 2-5 mechanical harvests of plantlets when the regenerative tissue is aged.

SUMMARY AND CONCLUSIONS

The continual-flow, liquid nutrient concept with the appropriate support matrices offers significant advantages over agar and conventional suspension culture for both enhanced growth and mechanization of plant micropropagation. The microporous polypropylene membrane achieves the aeration afforded by agar media yet allows optimal nutrient replenishment afforded by continual-flow liquid media. The trained growth of plantlets through the polypropylene netting provides a tool for mechanized mass handling of transfer, separation and singulation functions in plant tissue culture. Mechanized handling of plant tissue should be planned so that it enables a smooth interface with greenhouse and/or field handling of the crop.

LIST OF REFERENCES

Aitken-Christie, J. and Cathy Jones. 1987. Towards automation: Radiata pine shoot hedges *in vitro*. Plant Cell, Tissue and Organ Culture 8:185-196.

Aitken-Christie, J. and H. E. Davies. 1988. Development of a semi-automated micropropagation system. Proceedings of International Symposium on Protected Cultivation, Japan, May 12-15, 1988. Acta Horticulturae.

Cooke, R. C. 1979. Homogenization as an aid in tissue culture propagation of *Platycerium* and *Davallia*. HortScience 14(1):21-22.

Ferrell, M. A. 1987. Mega-Yield: artificial systemic interface system for plant micropropagation. Product Literature, Agro-Clonics, Inc., Orange, CA 92669.

Ford, C. F. 1989. Personal Communications. AgriStar, Inc., Sealy, TX 77474.

Fox, J. L. 1988. Plants thrive in ultrasonic nutrient mists. Bio/Technology 6:361.

Hamilton, R., H. Pederson and C. K. Chin. 1985. Plant tissue culture on membrane rafts. Bio Techniques 1:96.

Kong, Yan and Chee-Kok Chin. 1988. Culture of asparagus protoplasts on porous polypropylene membrane. Plant Cell Reports 7:67-69.

Levin, Robert. 1983. Process for plant tissue culture propagation. European Patent No. 0132414A2.

Levin, Robert. 1984. Plant tissue culture vessel. European Patent No. 0132413A2.

Levin, Robert, V. Gaba, B. Tal, S. Hirsh, D. DeNola and I. K. Vasil. 1988. Automated plant tissue culture for mass propagation. Bio/Technology 6:1035-1040.

Maene, L. J. and P. C. Debergh. 1985. Liquid medium additions to establish tissue cultures to improve elongation and rooting in vivo. Plant Cell, Tissue and Organ Culture 5:23-33.

Matsumoto, K. and H. Yamaguchi. 1989. Nonwoven materials as a supporting agent for *in vitro* culture of banana protocorm-like bodies. Tropical Agriculture 66(1):8-10.

Murashige, T. and F. Skoog. 1962. A revised media for rapid growth and bioassays with tobacco tissue culture. Physiol Plant. 15:473-497.

Simonton, W. and C. Robacker. 1988. Alternative system for micropropagation. ASAE Paper No. 88-1028. ASAE, St Joseph, MI, 49085-9659.

Sluis, C. J. and K. A. Walker. 1985. Commercialization of plant tissue culture propagation. IAPTC Newsletter 47:2-11.

Tisserat, Brent and C. E. Vandercook. 1985. Development of an automated plant culture system. Plant Cell, Tissue and Organ Culture 5:107-117.

Tisserat, Brent and C. E. Vandercook. 1986. Computerized long-term tissue culture for orchids. American Orchid Society Bulletin 55(1):35-42.

Vanderschaeghe, A. and P. C. Debergh. 1987. Automation of tissue culture manipulation in the final stages. Proceedings of Symposium on Vegetative Propagation on Woody Species, Pisa, Italy, September 3-5, 1987. Acta Horticulturae.

Weathers, P. J. and K. L. Giles. 1988. Regeneration of plants using nutrient mist culture. *In Vitro* Cellular and Developmental Biology 24(7):727-732.

Wochak, Z. S. 1979. Commercial tissue culture technology. American Nurseryman 149(2):10, 80-83.

Wood, Marcia. 1985. Tissue culture moves to the fast lane. American Vegetable Grower 33(11):70-71.

Young, R. E., S. A. Hale, N. D. Camper, R. J. Keese and J. W. Adelberg. 1989. An alternative, mechanized plant micropropagation approach. ASAE Paper No. 89-6092. ASAE, St. Joseph, MI, 49085-9659.

IN-SITU REAL TIME FLUOROMETRIC MEASUREMENT OF COENZYMES IN

MICROBIAL SYSTEMS

David P. Chynoweth and Michael W. Peck

Bioprocess Engineering Research Laboratory
Agricultural Engineering Department
University of Florida
Gainesville, Florida 32611

Several compounds associated with microbial activity
possess the property of fluorescence when exposed to
ultraviolet light. These compounds include, tyrosine,
tryptophan, phenylalanine, Vitamin A, flavins, ATP, ADP,
Vitamin B12, NAD(P)H, and Coenzyme F_{420}. The specificity of
their sensitivity and emitted fluorescence to narrow bands
permits their selective analysis in biological solutions.
Pyridine nucleotides (NADH/NADPH) are present at
approximately equivalent concentrations in all microorganisms
(1) and act as electron transferring coenzymes. In the
reduced form these molecules are fluorescent, while the
oxidized forms are essentially not fluorescent. Since the
excitation and fluoresced lights penetrate microbial cells,
their real-time in situ measurement in living microbial
cultures is possible without disturbance of the activities or
environment of the organisms. Measurement of NAD(P)H by
fluorometry therefore can provide an indication of the
oxidative state of the culture, overall microbial activity,
and numbers or quantity of microbial biomass. Fluorescence
monitoring probes developed commercially to measure NAD(P)H
have been used to monitor pure cultures of bacteria (2,3),
hybridoma cells (4), and yeasts (5,6). However, their use
with mixed-culture systems had not been reported until we
conducted a study of their suitability to monitor
methanogenic fermentations (7).

131

The methane fermentation (referred to as anaerobic digestion) is widely used for treatment of sewage sludge and industrial wastes and is nd en con sįder ation fo rconversion of solid wastes and biomass to methane. This fermentation is a multi-step process involving a consortium of microorganisms. Frequently these steps become uncoupled by differential responses to changes in feed, inhibitors, and environmental parameters such as temperature or oxidation/reduction potential. This may lead to instability in the fermentation and sometimes even a complete failure of the process. Techniques currently available to assess the real-time performance of anaerobic digesters (e.g. gas production, pH) are limited and by the time a response is detected, fermentation failure is frequently inevitable. Delays associated with off-line performance parameters (e.g. volatile acids and alkalinity) are unacceptable. There is a need for on-line analyses which provide a continuous real-time indicators of reactor performance and permit a control strategy to be employed. Since digester imbalance should be reflected in an interruption of normal electron flow in the metabolic processes of the methane fermentation, it should be possible to detect this imbalance through measurement of the concentrations of reduced electron coenzymes such as NAD(P)H. Another Coenzyme $_{420}$ is involved in electron transport and is characteristic of methanogenic bacteria and not likely to be present in other bacteria in anaerobic digesters (8). Measurement of concentrations of that coenzyme should therefore be related to the population size of methanogenic bacteria.

In this work, two fluorometric probes developed by BioChem Technology Corp. (Malvern, PA) were employed to measure activities of total and methanogenic microbial activities during deliberate overloading of a glucose-fed digesters. For both probes, the light source and detector are housed in a single probe which is inserted into a flow-through cell through which the contents of a glucose-fed digester are circulated. The NAD(P)H probe for total microbial activity has excitation light of 350 ± 40 nm and emission detector for emitted light of 460 ± 25 nm. The F_{420} probe for activity of methanogenic bacterial has excitation light of 406 ± 34 nm and emission detector for emitted light of 465 ± 20 nm. The oxidized rather than the reduced form of this coenzyme fluoresces. Fluorometric and conventional performance parameters were measured in a bench-scale (5-L) anaerobic digester operated at a hydraulic residence time of 24 days, temperature of 35°C, and increase in daily glucose feed rate from 8 to 64 grams to induce digester instability.

The NAD(P)H probe indicated that the fermentation was slightly imbalanced immediately following each daily feeding even at the lower loading rates; however both the probe and conventional measurements suggested that the digester recovered from these imbalances. By day 7, the NAD(P)H probe indicated severe imbalance, whereas other performance parameters (e.g. gas production and methane content, volatile acids, and pH) did not respond until day 8. Although the F_{420} probe also indicated instability on day 7, the response was less pronounced.

These results indicate that the NAD(P)H probe is a valid real-time indicator of the onset of digester instability as caused by digester overloading. Future experiments will evaluate other causes of instability such as feed inhibitors and relate extent of instability to measured probe response. Performance of the probe in the presence of particulate feedstocks will also be studied. Eventually the goal is to develop an on-line measurement and algorithm for control of the anaerobic digestion process using this measurement.

Acknowledgments:

This work was sponsored by a co-funded program with the Gas Research Institute and the Institute of Food and Agricultural Science, University of Florida.

References:

1. London , J. and Knight, M., "Concentrations of nicotinamide dinucleotide coenzymes in microorganisms," J. General Microbiol., 44, 241-254, (1966).

2. Luong, J. H. T. and Carrier, D. J., "On-line measurement of culture fluorescence during cultivation of Methylomonas mucosa," Appl. Microbiol. Biotechnol., 24, 65-76, (1986).

3. Rao, G. and Mutharasan, R., "NADH levels and solventogenesis in Clostridium acetobutylicum: new sights through culture fluorescence," Appl. Microbiol. Biotechnol., 30, 59-66, (1989).

4. Armiger, W. B., Forro, J. F., Lee, J. F., MacMichael, G., and Mutharasan, B., "On-line measurement of hybridomas growth by culture fluorescence," Pro. BioExp '86, Boston, MA, (1986).

5. Meyer, C. and Beyeler, W., "Control strategies for continuous bioprocesses based on biological activities," Biotechnol. Bioeng., 26, 916-924, (1984).

6. Armiger, W. B., Forro, J. F., Lee, J. L., Montalvo, L. M., and Zabriskie, D. W., "The interpretation of on-line process measurements of intracellular NADH in fermentation processes," Chem. Eng. Commun., 45, 197-206, (1986).

7. Peck, M. W. and Chynoweth, D. P., "On-line monitoring of the methanogenic fermentation by measurement of culture fluorescence," Biotechnol. Let., 12, 17-22, (1990).

8. Peck, M. W. and Archer, D. B., "Methanogens and methods for their enumeration," Internat. Indust. Biotechnol., 9, 5-12, (1989).

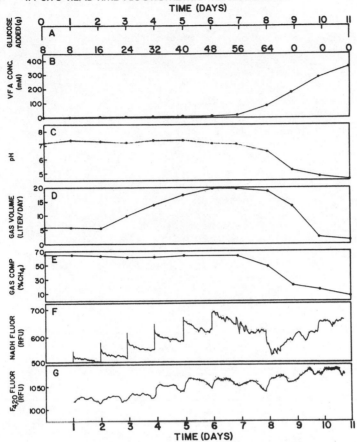

Figure 1.   On-line fluorescence and off-line measurements of glucose-fed anaerobic digester subject to overloading (with permission from reference 7).

POLICY ISSUES IN BIOTECHNOLOGY

John F. Gerber and Alvin L. Young

Office of Agricultural Biotechnology,

United States Department of Agriculture

Washington, D.C. 20250

Plant and animal breeding and strain selection of microorganisms have been classical tools used in agriculture for animal, plant and microorganism improvement. Exactly how this was originally done is not at all clear, but clearly Gregor Mendel changed not only the way plant and animal breeding was done, but also the way it was viewed and used. In the same way the discoveries that genes were made up of repeated and well defined chemical units that were subject to manipulations again changed the way animal, plant and microorganism improvement will be done as well as the way it is viewed.

Since the beginning of the world man has been largely dependent upon fortuitous events such as mutations and selection to improve microorganisms used for food processing and or production of chemicals and medicinals. Today, with the new molecular "tools" of biology these organisms can be engineered or designed to produce specific products or perform specific tasks.

There is and will be continuing debate over whether this new technology is good or bad. Opinions will differ, but the reality is that the way breeding and microbial improvement will be done has forever changed. It means that in the future the use

137

of domesticated plants, animals and microorganisms to produce food, fiber, and ornamental plants, in short agriculture, will be different. The United States Department of Agriculture (USDA) is the primary Department of the Federal government charged with the welfare of agriculture and the food and fiber system. USDA has responsibility for the development and implementation of policies for agricultural biotechnology. What are the biotechnology policy issues of concern to the USDA?

*   Development of safe, understandable, workable rules for field research and commercialization of genetically modified plants, animals and microorganisms.

*   Insuring that the biotechnology discoveries and inventions from the UDSA laboratories, or produced with funds provided by USDA, are made available to users through commercialization.

*   Using the tools of biotechnology to keep the U.S. and especially U.S. agriculture competitive in a global economy.

*   Providing funds and leadership to insure that agricultural scientists can develop and use the tools of biotechnology to benefit the agricultural producer, food processor, consumer and society.

*   Insuring that the U.S. maintains a high quality, well motivated community of agricultural research scientists.

*   Using biotechnology to reduce dependence on traditional pesticides, and to reduce or change cultural practices and use of purchased inputs that are harmful to water and environmental quality.

*   Using the tools of biotechnology to improve the quality and healthfulness of our food supply.

*   Harmonizing   biotechnology   regulatory
     policies between and among our trading
     partners.

*   Providing information about the uses and
     implications of biotechnology to farmers,
     consumers, processors and the general
     public.

*   And above all, doing all of this in a safe,
     responsible affordable way.

     Since the uses of biotechnology are ubiquitous,
it is important that there be coordination and
cooperation within the Department on crucial issues
related to biotechnology.   The Office of
Agricultural Biotechnology was established within
the Office of the Secretary of Agriculture for this
purpose. Its role is not programmatic, but to
identify policy issues and assist in the process of
policy development.

SAFE WORKABLE RULES

     The lack of clear, understandable, workable
rules for biotechnology is probably the greatest
impediment to the development of agricultural
biotechnology today. While the Animal and Plant
Health Inspection Service (APHIS) has issued
approximately 100 permits for field research with
genetically altered organisms, there are no permits
yet issued for the commercial field use of
genetically modified plants, animals and
microorganisms. The APHIS efforts are commendable
because without them it would not be possible to
conduct field research in the U.S., but they are
not we believe the final or best solution.

     The USDA under the leadership of the Office of
Agricultural Biotechnology with the advice and
counsel of The Agricultural Biotechnology Research
Committee (ABRAC) has been developing Guidelines for
Research with Genetically Modified Organisms Outside

Contained Facilities (Guidelines). This effort reached a significant milestone in April 1990 when the Committee gave their approval to the Guidelines as the basis for the development of an environmental impact statement (EIS) as required by the National Environmental Policy Act (NEPA). A series of scoping or public hearings will be conducted throughout the nation to identify issues and alternatives which need consideration. Once these sessions are completed, the EIS will be drafted and published. Following a period during which comments will be received the final EIS will be prepared and the Guidelines implemented. This process may take 2 years.

When the Guidelines are completed, it is anticipated that the Institutional Biosafety Committees of universities, colleges and industry will review all proposed research covered by the Guidelines. They may approve that which is clearly safe and refer to the USDA for further review those research projects which have broad implications to human health or to the environment including natural and managed ecosystems. The proposed scheme is similar in many ways to the process used by the National Institutes of Health (NIH). Without going into further details, it is clear that the USDA is developing workable rules for field research that is essential if the new applications of the technology are to continue.

BIOTECHNOLOGY IMPROVING COMPETITIVENESS

What is the USDA doing to shape policy so that the fruits of publicly funded research are made available to the public?

A little history may be in order. Before World War II only the USDA provided Federal funds for research. In part because science played such a big role in winning the war, it became post war science policy to fund research. USDA funded research was tied to the use of the Cooperative Extension Service to assist farmers in adapting and using new technology. The funding provided by other Federal agencies didn't have such a connection, and as a result U.S. science and scientists became the patrons of Federal funds rather than constituents such as farmers. These funds for research grew at an annual rate of about 15% during the 1950's and 1960's. Any scientist that wasn't a patron of this system was foolish and improvised. The problem was that the Federal government was only the mechanism by which the public provided funds for research, not the user or customer. Somehow this was forgotten, and research was designed and structured in ways that would insure a continued flow of funds and bring more scientists into the system. In this manner the U.S. built the most powerful scientific research machine in the world. But during the 1970's and the 1980's the rate of increase in research funds begin to decline and at the same time the growth of U.S. industrial output began to decline. Countries that were spending much less on research than the U.S. were outstripping the U.S. in rate of industrial growth. It appeared that they were using the U.S. research machine to fuel the development of technology that in many cases was clearly superior to the U.S. technology.

At issue now is the regaining of U.S. technological competitiveness by recoupling industry and research via technology transfer. This recoupling has the added advantage of making the fruits of biotechnology research available to the public. Thus, remaining competitive and making sure the public benefits from publicly funded research have become inexorably entwined. Both Federal and State governments have taken actions to assist this process. Some of the things that were done:

* Transferred ownership of patents to university and Federal labratories,

* Allowed scientists to receive extra benefits as royalties and license fees,

* Permitted scientists/inventors to become principals in companies, especially

important since so many biotechnology
companies are startup companies,

*   Established the Small Business Innovative
     Research Program. (many states enacted
     companion programs),

*   Provided tax incentives to companies that
     made additional investments in research,
     and

*   Provided more patent examiners to deal
     with biotechnology patents.

All of this was done to encourage scientists
and engineers to become proactively engaged in
transforming basic science into useful products, and
thereby improving and maintaining our
competitiveness and making the fruits of science
available to consumers.

Discoveries and inventions from biotechnology
only become available to the user when private
companies produce and market them. Therefore, it is
appropiate USDA policy to encourage the transfer of
inventions and discoveries from the Agricultural
Research Service (ARS) and from funds provided to
the state universities and land grant colleges
through the Cooperative State Research Service
(CSRS) to the private sector. The Department uses
the tools provided by Federal actions and thus,
participates in the SBIR program, has decentralized
the ownership and benefits from patents to the
inventors and laboratories, and has developed a
proactive program to assist industry in the
commercialization of biotechnology patents and
inventions. USDA has also transferred the patent
rights and benefits from inventions and discoveries
made with CSRS funds to the institutions that
receive these funds. In addition, USDA has
participated in consortia, institutes and provided
funds to universities and colleges which will be
used in programs that are designed to assist and
expedite the transfer of biotechnology to the
private sector.

PROVIDING RESEARCH FUNDS AND LEADERSHIP

Agricultural biotechnology research has lagged behind biomedical biotechnology due in part to the great disparity in Federal funding for human health research through NIH as compared to Federal funds for agricultural research. The disparity between animal and plant biotechnology is even worse because animal biotechnology research has benefitted from NIH funding, but plant biotechnology has been funded at the Federal level almost exclusively by USDA, the National Science Foundation (NSF) and the Department of Energy (DOE). Several billion dollars are spent annually on biotechnology research by NIH and at best, a few hundred million by USDA, NSF and DOE. That is why the Department has placed such a high priority on research funding and the new National Research Initiative which could, over the next few years, result in several hundred million dollars annually for new research. Many of these new funds will be used for plant, microbial, arthropod, and animal biotechnology. In addition, USDA is encouraging the research community to develop genetic maps and sequences of plant and animal genomes. This may appear as overly ambitious when compared to the projected cost of the human genome project, but the Department will interact closely with the human genome project so that the spinoffs from this project will be available to the agricultural sciences.

MAINTAINING A COMMUNITY OF AGRICULTURAL RESEARCH SCIENTISTS

We are in danger of running out of qualified agricultural scientists in the U.S. This shortage will not only continue but will worsen because there are fewer students in high school and colleges interested in the biological sciences, let alone agriculture! There is a need for more scientist trained at the Ph.D. level and this will only happen if young men and women demonstrate an aptitude and interest in science. The Department of Agriculture has initiated a variety of programs to encourage more students to pursue careers in science,e.g., The Ag in the Classroom program, and to encourage graduate students, by providing funds to universities to support graduate fellows in the agricultural sciences.

USING BIOTECHNOLOGY TO
REDUCED DEPENDENCE ON CHEMICALS

Biotechnology may have its first real impact on agriculture by replacing some pesticides and fumigants with materials of biological origin or with living organisms. The reason for success is the long history of scientific research in biological control of pests with other organisms. Since the mode of action of some of these methods is known, it is possible to genetically engineer the genes for a toxic protein from the microorganism <u>Bacillus thuringiensis</u>, directly into the plant. When the plant makes its own pesticide very little gets into the water, and the total amount required is much less because the protection is available when and where it is need. It is also possible to make killed organisms which contain pesticidal activity and to find other means to make the plant or animal more resistant or tolerant of insects, nematodes, and diseases. Clearly, this strategy will improve water and air quality and reduce residues of pesticides in the soil or on plants.

Is the food safe for human consumption from such plants or when biopesticides are used? The scientific community feels that they are safe based on what is known about the compounds. Yet the consumer wants to know that the risks are negligible. USDA is concerned and committed to providing assurance that the food supply is safe.

Herbicide tolerance can be transferred to plants so that weeds may be more easily controlled. Tolerance to selective herbicides could be enhanced with biotechnology, but there are widely differing opinions about the desirability of such action. The policy of the USDA is to let the science and scientists provide the information through research that will form the basis of decision, and not to prejudge or prevent research a priori. If herbicides can make minimum tillage successful, with a subsequent reduction in runoff and erosion, then this is a substantial environmental benefit, and efforts to achieve this should be pursued.

IMPROVING THE QUALITY OF THE FOOD SUPPLY

The USDA came into existence when most of the population of the U.S. was involved in agriculture. To that degree it was the "people's" Department. Consumers still look to the Department for assurance of a safe, dependable, economical supply of food. Biotechnology can assist in this effort in ways that were not possible only a few years ago. For example, antisense genes have been successfully transferred to tomatoes, that permit the tomato to ripen and remain ripe without softening and deteriorating. This is just the beginning of an effort by scientists to improve the tastiness, the nutritional quality, shelf life of foods by delaying spoilage, and removing certain naturally occurring but undesirable components. These components may be excess fats or cholesterol, fatty acids, alkaloids, or other toxins.

Monitoring of the food supply to detect contamination is a costly and time consuming processes. There are never enough resources to accomplish the task. Through the use of biotechnology it is possible to devise quick, accurate tests that will detect contaminants or toxins, and perhaps verify that processed foods contain components as labeled. Such test would not only greatly simplify the Departments efforts, but would also provide the producer, processor and retailer a rapid simple method to monitor their products as they move through the market chain.

HARMONIZING BIOTECHNOLOGY REGULATIONS

Today we are in a global economy with trade following new and different patterns. Rarely is a product produced solely for domestic consumption. Patterns of regulation developed in the past have sometimes become trade impediments and are difficult to modify. We have a chance with this new technology to develop harmonization with our trading partners so that the basis for rules and regulations are scientifically sound and well understood. Such rules will encourage, not inhibit trade. The USDA is in constant dialogue with our European and Pacific trading partners about the regulations which we are proposing. Undoubtedly we will have differences, but we hope to arrive at a common basis for commerce and trade. The task of harmonizing is slow, tedious, but most important, it can be very fruitful.

PROVIDING INFORMATION ABOUT BIOTECHNOLOGY

Democracies function best when the electorate is well informed. It is a USDA policy to supply information about agricultural biotechnology to farmers, processors, industry, consumers and the public. There are several ways which are being used to accomplish this task. All the agencies of the USDA have informational programs to inform their constituents about their programs including biotechnology. The Office of Agricultural Biotechnology publishes <u>Biotechnology Notes</u> monthly to provide better communication within the Department. The National Biological Impact Assessment Program (NBIAP) maintains the Agricultural/Environmental Biotechnology Bulletin Board. The National Agricultural Library contains most of the pertinent scientific documentation about biotechnology which is used by a large audience outside the USDA. In addition, representatives from all branches of the USDA participate in scientific meetings, public hearings, and with the press about biotechnology.

IV.

NON-INVASIVE DETECTION OF TISSUE
AND ORGAN SYSTEM FAILURE: IMAGING

QUANTITATIVE EVALUATION OF TWO SPECT RECONSTRUCTION TECHNIQUES

[1]Ballard, J.G., [1,2]Tsui, B.M.W., and [2]Johnston, R.E.

[1]Curriculum in Biomedical Engineering, [2]Department of Radiology

University of North Carolina, Chapel Hill, N.C.

ABSTRACT

Accuracy in the measurement of radioactivity in vitro was used to evaluate SPECT images reconstructed by two methods, the expectation maximization (EM) algorithm for the maximum likelihood (ML) estimates and a modified Chang algorithm. Both reconstruction methods use a measured attenuation map to compensate for attenuation in the patient. Scatter compensation was accomplished by a windowed subtraction method using data collected from two energy windows. Two phantoms with different configurations and attenuating properties were used in the evaluation. Concentration estimates from both phantoms were estimated with less than 10% error. Both reconstruction techniques performed equally well with regards to quantitation. The ML-EM method approached it's final estimate in 15 to 20 iterations while the Chang method varied little after the first iteration. However the ML-EM provided better image quality in terms of noise fluctuations.

INTRODUCTION

Single photon emission computed tomography (SPECT) is a medical imaging technique that provides a variety of physiologic and functional information about the patient. SPECT images represent the distribution of a radiopharmaceutical inside a patient's body. Typically, photons emitted from the radiopharmaceutical are acquired as projection images using a scintillation camera. SPECT projection images collected from different angles around the patient are used to reconstruct the 3-D distribution of radioactivity.

The accuracy of the reconstructed images is dependent on the reconstruction algorithm and methods used to compensate for physical factors such as photon attenuation,detector response and scatter. Other factors such as system dead-time may also significantly affect the quantitative information.

This study was designed to evaluate the relative merit of the iterative ML-EM and Chang reconstruction methods with regards to quantitation [1, 2, 9, 12]. Reconstruction methods have traditionally been evaluated based on image quality or lesion detectability. Because quantitative evaluation of images is becoming more common the accuracy of quantitative information in the image needs to be assessed.

The iterative ML-EM and Chang methods were chosen because they allow use of a measured attenuation map to compensate for non-uniform attenuation. Conventional compensation and reconstruction methods assume a uniform attenuation distribution in the patient.Uniform attenuation compensation does not accurately compensate for attenuation in non-uniform attenuating regions of the body such as the chest. The ML-EM algorithm produces images with good noise characteristics but convergence to a final image is slow. Chang arrives at a final image quickly but at the expense of larger noise fluctuations which may interfere with evaluation of the image.

MATERIALS AND METHODS

Phantoms were used to simulate the body of a patient. Phantoms also had the advantage that the absolute concentration of radionuclide inside the phantom was accurately known. The uniform attenuating phantom had a circular cross-section with a diameter of 23cm. A 6cm diameter hollow sphere positioned in the center. The large sphere was used to minimize the effect of detector response on the estimates of concentration inside the sphere. The non-uniform attenuating phantom is designed to simulate the chest region of a patient. It had an elliptical cross-section with major and minor axis equal to 31cm and 23cm respectively. Bottles filled with Styrofoam beads and water to simulate the attenuation of 'lungs' were added to provide non-uniform attenuation [12]. A 6cm diameter hollow sphere was also positioned in the center of this phantom. Each phantom was filled with water. Technetium-99m (Tc-99m) was added to the inside of the sphere and into the surrounding volume of water to a concentration ratio of 10 to 1.

A GE 400 AT gamma camera was used to collect all experimental data into 64x64 element arrays. Projection data was collected from two energy windows, the 'scatter' energy window was set for 90-123 kev and the photopeak energy window was set for 124-156 kev. The acquisition of transmission projection data to determine an attenuation map is described elsewhere [5, 12]. A point source of known activity was imaged to determine the sensitivity calibration factor in counts/μCi/sec for the imaging system.

Projection data from both phantoms were reconstructed by the iterative ML-EM and Chang reconstruction methods. The attenuation map was incorporated in the projector and backprojector of the iterative algorithm to compensate for attenuation [1, 5]. The windowed subtraction method was used to compensate for scatter [3, 4, 6, 7, 8, 13]. In this method a fraction k of the scatter projection data is subtracted from the photopeak projection data. The k value used was 0.5 which had been determined by Monte Carlo simulations of a similar experimental situation [3, 7, 10]. Dead-time compensation was carried out assuming a paralyzable model with a dead-time value of 0.49 μsec [11]. Decay compensation for each frame of the acquired data followed dead-time compensation.

A transaxial slice through the center of the sphere was reconstructed using the compensation methods described above. Regions of interest were placed over the sphere and background of the reconstructed images and the detected counts within the regions were averaged. These counts, the pixel size calibration of the images and the sensitivity of the imaging system were used to estimate the concentration of Tc-99m in the phantom. Results are presented as percentage error in the concentration estimates to allow comparison between experiments with slightly different concentration values.

RESULTS

The percent error for estimating the concentration of Tc-99m in the sphere of both phantoms using the iterative ML-EM reconstruction method with attenuation compensation is presented in Figure 1(a). The percent error

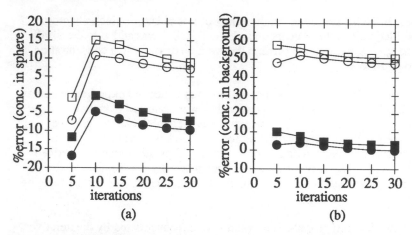

Figure 1. ML-EM reconstruction with attenuation compensation. (a).Percent error in estimating the concentration of Tc-99m inside a 6cm diameter sphere in a uniform attenuating circular cross-section phantom and a non-uniform attenuating elliptical cross-section phantom. (b). Percent error in estimating the background. (O) uniform attenuating phantom without scatter compensation, (●) uniform attenuating phantom with scatter compensation, (□) non-uniform attenuating phantom without scatter compensation, (■) non-uniform attenuating phantom with scatter compensation

of the sphere changes dramatically in the first 10 iterations. Scatter compensation has improved the quantitative accuracyfor the uniform attenuating phantom and the non-uniform attenuating phantom. The phantoms have a percent error of about -7% after scatter compensation. Figure 1(b) shows the accuracy of the background estimates. Scatter compensation improves the concentration estimates for both the uniform and non-uniform attenuating phantoms to about 5%. Also the relative change of the percent error in the first 10 iterations for the background is smaller than for the concentration estimates inside the sphere. The percent error did not change significantly after 15 iterations.

The percent error, as shown in Figure 2, with the modified Chang attenuation compensation method did not change much in the 5 iterations

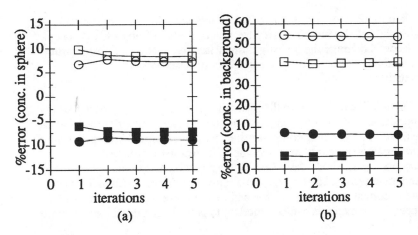

Figure 2. Chang attenuation compensation method. (a).Percent error in estimating the concentration of Tc-99m inside a 6cm diameter sphere in a uniform attenuating circular cross-section phantom and a non-uniform attenuating elliptical cross-section phantom. (b). Percent error in estimating the background. (O) uniform attenuating phantom without scatter compensation, (●) uniform attenuating phantom with scatter compensation, (□) non-uniform attenuating phantom without scatter compensation, (■) non-uniform attenuating phantom with scatter compensation

that were evaluated. After scatter compensation the concentration estimate inside the sphere of both the uniform and non-uniform attenuating phatom was less than 10%, Figure 2(a). The background concentration estimates of both phantoms is shown in Figure 2(b) to have approximately 5% error with the Chang method.

DISCUSSION AND SUMMARY

Quantitative comparison of the results from the iterative ML-EM and Chang reconstruction methods with attenuation compensation indicate that both methods had similar accuracy in estimating the radionuclide concentration for both the uniform and non-uniform attenuating media. The Chang method arrives at an acceptable solution to the reconstruction problem much sooner than the ML-EM method. However based on image quality the ML-EM method may be preferred over Chang [8, 12]. The results do encourage the investigation of iterative methods to reconstruct quantitatively accurate images, especially in cases with non-uniform attenuating medium.

The relation of scatter to the source location and the geometry and composition of the attenuating medium complicates scatter compensation. The use of better scatter compensation techniques should provide further improvements in quantitative accuracy.The reconstruction and compensation methods described in this paper have provided good quantitative results. However, evaluation with different attenuation and source geometries must be performed before these reconstruction and compensation methods can be relied on to provide quantitatively accurate results.

In conclusion, the ML-EM and Chang reconstruction methods with attenuation and scatter compensation estimated the concentration of radionuclide in each phantom equally well. Concentration estimates of the 6cm diameter sphere and background of both phantoms were within 10% error. More research is necessary to develop better compensation procedures for attenuation, scatter, detector response, and dead-time to further improve the quantitative accuracy.

BIBLIOGRAPHY

1.   Budinger, T. F., G. T. Gullberg and R. H. Huesman. "Emission Computed Tomography." Image Reconstruction From Projections: Implementation and Applications. Herman ed. 1979 Springer-Verlag. New York.

2.   Chang, L.-T. "A Method For Attenuation Correction in Radionuclide Computed Tomography." IEEE Trans. Nucl. Science. NS-25(1): 638-643, 1978.

3.   Floyd, C. E., R. J. Jaszczak, C. C. Harris, K. L. Greer and R. E. Coleman. "Monte Carlo evaluation of Compton scatter subtraction in single photon emission computed tomography." Med.Phys. 12(6): 776-778, 1985.

4.   Gilardi, M. C., V. Bettinardi, A. Todd-Pokropek, L. Milanesi and F. Fazio. "Assessment and Comparison of Three Scatter Correction Techniques in Single Photon Emission Computed Tomography." J.Nucl.Med. **29**(12): 1971-1979, 1988.

5.   Gullberg, G. T., R. H. Huesman, J. A. Malko, N. J. Pelc and T. F. Budinger. "An Attenuated Projector-Backprojector for Iterative SPECT Reconstruction." Phys.Med.Biol. **30**(8): 799-816, 1985.

6.   Jaszczak, R. J., C. E. Floyd Jr. and R. E. Coleman. "Scatter Compensation Techniques For SPECT." IEEE Trans.Nucl.Science. **NS-32**(1): 786-793, 1985.

7.   Jaszczak, R. J., K. L. Greer, C. E. Floyd Jr., C. C. Harris and R. E. Coleman. "Improved SPECT Quantitation Using Compensation for Scattered Photons." J.Nucl.Med. **25**(8): 893-900, 1984.

8.   Koral, K. F., F. M. Swailem, S. Buchbinder, N. H. Clinthorne, W. L. Rogers and B. M. W. Tsui. "SPECT Dual-Energy-Window Compton Correction: Scatter Multiplier Required for Quantification." J.Nucl.Med. **31**(1): 90-98, 1990.

9.   Lange, K. and R. Carson. "EM Reconstruction Algorithms for Emission and Transmission Tomography." J.Comp.Asst.Tomo. **8**(2): 306-316, 1984.

10.   Manglos, S. H., C. E. Floyd, R. J. Jaszczak, K. L. Greer, C. C. Harris and R. E. Coleman. "Experimentally Measured Scatter Fractions and Energy Spectra as a Test of Monte Carlo Simulations." Phys.Med.Biol. **32**(3): 335-343, 1987.

11.   Perdikaris, N. Evaluation of a Rotating-Camera System For Single Photon Computed Tomography. 1984.

12.   Tsui, B. M. W., G. T. Gullberg, E. R. Edgerton, J. G. Ballard, J. R. Perry, W. H. McCartney and J. Berg. "Correction of Nonuniform Attenuation in Cardiac SPECT Imaging." J.Nucl.Med. **30**(4): 497-507, 1989.

13.   Yanch, J. C., M. A. Flower and S. Webb. "A Comparison of Deconvolution and Windowed Subtraction Techniques for Scatter Compensation in SPECT." IEEE.Trans.Med.Imaging. **7**(1): 13-20, 1988.

MATCHING WITH BOUNDARY CONSTRAINED MODEL FOR LEFT VENTRICULAR CONTOUR RECOGNITION*

N. Fan[#+], C.C. Li[#], B.G. Denys, MD[+], and P.S. Reddy, MD[+]

[#]Department of Electrical Engineering and
[+]Department of Medicine
University of Pittsburgh
Pittsburgh, PA 15261

Abstract

A model-based approach for automatic recognition of left ventricular contour in a cineangiogram has been developed. This paper presents an adaptive matching algorithm in the process to register a boundary constrained model to the left ventricle location in a X-ray image. This method has been experimented on cineangiograms of 30 subjects with reasonable success.

INTRODUCTION

The accurate determination of left ventricular border in cineangiograms is a fundamental requirement for the image-based cardiac function assessment. It is also the first step for 3-D surface reconstruction of left ventricle based on only two orthogonal views [1]. Currently, manual tracing is employed routinely to obtain the necessary information such as volume, filling rate, and so on.

Computer determination of left ventricular border has been a difficult problem. Due to the regional low contrast of the X-ray image, segmentation based on the gray scale thresholding or the conventional edge detection method often produces poor results, such as broken regions or

---

*This work is partially supported by a grant from Siemens Gammasonics Inc.

disconnected edges. Recent endeavours of incorporating multiple templates and knowledge base [2, 3] have led to considerable improvements. However, because the templates are to be manually aligned to the left ventricle image, these methods suffer from the following two problems: *i*)the process is not automatic so that it cannot be done in batch; and *ii*)the results are rarely reproducible due to the inconsistency in each manual alignment. The deformable model-based object recognition [4, 5, 6, 7] has recieved much attention in the computer vision field recently. In this approach, a template is required to be placed near the object (manually), it then deforms itself to match the object (automatically). Although this method improves the result of the pure manual alignment, the problems remain.

This paper is to present an approach which applies a boundary constrained model [8] to automatically extract the left ventricle in a segmented binary image. Because the boundary constrained model specifies the shape deformation in expansion and contraction without introducing additional parameters, the matching parameter space is smaller than those of other deformable models. Therefore, multiple initial solutions can be tried and the globally best matched final solution can be obtained as the correct position of the left ventricle.

The entire recognition process consists of three stages: segmentation, model matching, and contour determination. This paper will address the issue of the second stage. The segmentation is performed by extracting the significant grey scale changes both within a single image and between two consecutive images during the heart contraction (or expansion). An adaptive matching algorithm is used to register the boundary constrained model to the left ventricle location in each image. The ventricular contour is then determined within the region confined by the matched model via a dynamic programming process.

THE BOUNDARY CONSTRAINED MODEL

A beating heart is an object which contracts and expands between two limits, the highest volume shell at the end of diastole, and the smallest volume shell at the end of systole. The 2-d projection of a left ventricle in a cineangiogram varies correspondingly within a 2-d confined region. A pair of such confined regions for left ventricle in RAO and LAO projections, based on the data of 30 subjects, is shown in Figure 1. In any instance, the border of a projected left ventricle lies within its respective confined region.

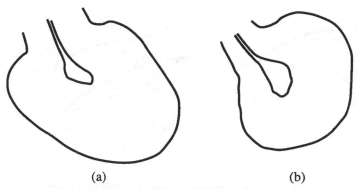

(a)   (b)

Figure 1. Two confined regions of a projected left ventricle:
(a) at right anterior oblique (RAO) angle;
(b) at left anterior oblique (LAO) angle.

The boundary constrained model specifies the confined region in terms of its medial involution chords. Assume that the medial axis of the confined region is a single curve without branches. Then, the symmetric axis transform [9] of the confined region can be written as $x = x(s)$; $y = y(s)$; $R = R(s)$. A medial involution chord [10] is a line segment between two points $\{p, q\}$ at which a maximal disc contacts the border of the confined region, as shown in Figure 2(a); they are given by [8]

$$p : (x, y) + Re^{\, j\left(\alpha+\beta+\frac{\pi}{2}\right)}, \quad q : (x, y) + Re^{\, j\left(\alpha-\beta-\frac{\pi}{2}\right)} \tag{1}$$

For simple contacts,

$$\alpha = \tan^{-1}\frac{dy}{dx}, \quad \beta = \sin^{-1}\frac{dR}{ds}; \tag{2}$$

and for finite contacts (contacts with a contiguous piece of arc),

$$\alpha = (1-\tau)\alpha_- + \tau\alpha_+, \quad \beta = (1-\tau)\beta_- + \tau\beta_+, \tag{3}$$

where $\tau \in (0, 1)$, and

$$\alpha_\pm = \lim_{\substack{\delta \to 0 \\ \delta > 0}} \tan^{-1}\frac{d}{dx}y(s\pm\delta), \quad \beta_\pm = \lim_{\substack{\delta \to 0 \\ \delta > 0}} \sin^{-1}\frac{d}{ds}R(s\pm\delta).$$

The medial involution partitions the border of the confined region into two seperated curves: the interior border B_I and the exterior border B_E. It can be

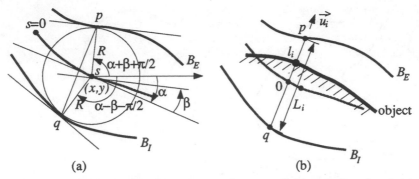

Figure 2. Schematic diagrams illustrating a medial involution chord: (a) definition; (b) intersection with an object boundary.

proven that for every point inside the confined region there is a unique medial involution chord passing through this point, and two medial involution chords can never cross each other [8]. The model template consists of a discrete set of m medial involution chords, as shown in Figure 3.

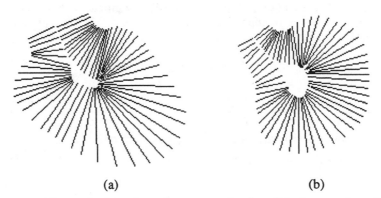

Figure 3. Two boundary constrained models for a projected left ventricle: (a) at RAO angle; (b) at LAO angle.

Along each chord, a 1-d coordinate system is set up with its origin O at the skeleton point and the positive direction towards the exterior border, as shown in Figure 2(b). For an object lying inside the confined region with its boundary intersecting the ith chord at l_i, while the length of the ith chord is L_i, we define the *deformity* d and the *roughness* r as follows:

$$d = \frac{1}{m}\sum_{i=0}^{m-1} \left(\frac{l_i}{L_i} - \bar{l}_{avg}\right)^2, \qquad where \quad \bar{l}_{avg} = \frac{1}{m}\sum_{i=0}^{m-1} \frac{l_i}{L_i} \tag{4}$$

$$r = \frac{1}{m}\sum_{i=1}^{m-1} \left(\frac{l_i}{L_i} - \frac{l_{i-1}}{L_{i-1}}\right)^2 \tag{5}$$

Because d and r vary with respect to different alignments between the model template and the object, the one with the minimum d is defined as the matched alignment. The corresponding values d_o and r_o indicate the object deformity and roughness respectively.

From (4), a family of contours inside the confined region with $\frac{l_i}{L_i} = \bar{l}_{avg}$ $\forall i$ has zero deformity. We call them the *prototypes*, which mimic an ideal left ventricle contracting and expanding in propotion within the confined region. Because a real left ventricle is different from the ideal model, small degrees of deformity and roughness should be tolerated. Therefore, threshold values d_m and r_m are to be used. In other words, if and only if $d_o \le d_m$ and $r_o \le r_m$, the object is matched and recognized as a left ventricle.

MATCHING

Since the cineangiogram is generated from the parallel projection of X-rays, the scale change of the object should be zero. Thus, the matching between the boundary constrained model and the projected left ventricle image is a congruence transformation. The task of the matching is then to find the correct values of the transformation parameters: θ, x_o, and y_o such that the deformity d is minimized.

Since the congruence transformation contains trigonometric nonlinearities, we first linearize it by assuming a small change in the parameters: $\delta\theta$, δx_o, and δy_o, at each step. Let us also assume that the object boundary point intersecting with the ith chord of the model, after a small movement of the model, lies on the tangent line of the boundary at the current intersecting point, as shown in Figure 4. Let $l_{avg,i} = \bar{l}_{avg} L_i$ and

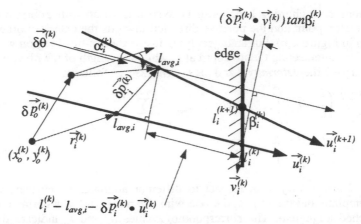

Figure 4. The geometry in which one chord of a boundary constrained model intersects with a binary image edge.

recognize that $l_{avg,i}^{(k+1)} \approx l_{avg,i}^{(k)} = l_{avg,i}$, then at the kth step [8],

$$l_i^{(k+1)} - l_{avg,i} = \frac{\cos\beta_i}{\cos(\alpha_i - \beta_i)} \{ \ [\ l_i^{(k)} - l_{avg,i} - \delta\vec{p}_i^{(k)} \bullet \vec{u}_i^{(k)} \ ]$$

$$+ (\delta\vec{p}_i^{(k)} \bullet \vec{v}_i^{(k)}) \tan\beta_i \} \qquad (6)$$

Because the multiplication factor and the β_i in the second term on the right side are difficult to compute, instead of minimizing $d^{(k+1)}$, we simplify our objective function in each step as

$$L^{(k)} = \sum_{i=0}^{m-1} \frac{1}{L_i^2} (\ \delta l_i^{(k)} - \delta\vec{p}_i^{(k)} \bullet \vec{u}_i^{(k)} \ )^2, \quad where \ \ \delta l_i^{(k)} = l_i^{(k)} - l_{avg,i} \qquad (7)$$

For congruence transform of a small variation $\delta\vec{p}_i \approx \delta\vec{p}_o + \vec{\delta\theta} \times \vec{r}_i$,

$$L = \sum_{i=0}^{m-1} \frac{1}{L_i^2} [\ \delta l_i - (\delta x_o - \delta\theta r_{iy})u_{ix} - (\delta y_o + \delta\theta r_{ix})u_{iy} \ ]^2 \qquad (8)$$

Differentiate L with respect to δx_o, δy_o and $\delta\theta$ and set the results to zero, we have

$$\delta x_o (\sum_{i=0}^{m-1} \frac{u_{ix}^2}{L_i^2}) + \delta y_o (\sum_{i=0}^{m-1} \frac{u_{ix}u_{iy}}{L_i^2}) + \delta\theta (\sum_{i=0}^{m-1} \frac{t_i u_{ix}}{L_i^2}) = \sum_{i=0}^{m-1} \frac{\delta l_i u_{ix}}{L_i^2}$$

$$\delta x_o (\sum_{i=0}^{m-1} \frac{u_{ix}u_{iy}}{L_i^2}) + \delta y_o (\sum_{i=0}^{m-1} \frac{u_{iy}^2}{L_i^2}) + \delta\theta (\sum_{i=0}^{m-1} \frac{t_i u_{iy}}{L_i^2}) = \sum_{i=0}^{m-1} \frac{\delta l_i u_{iy}}{L_i^2} \qquad (9)$$

$$\delta x_o (\sum_{i=0}^{m-1} \frac{t_i u_{ix}}{L_i^2}) + \delta y_o (\sum_{i=0}^{m-1} \frac{t_i u_{iy}}{L_i^2}) + \delta\theta (\sum_{i=0}^{m-1} \frac{t_i^2}{L_i^2}) = \sum_{i=0}^{m-1} \frac{\delta l_i t_i}{L_i^2}$$

where $t_i = r_{ix}u_{iy} - r_{iy}u_{ix}$. Eq. (9) can be solved by the SVD method [11] to avoid singularity problem. Then, the transformation parameters are updated by $x_o^{(k+1)} = x_o^{(k)} + \delta x_o$, $y_o^{(k+1)} = y_o^{(k)} + \delta y_o$, and $\theta^{(k+1)} = \theta^{(k)} + \delta\theta$ until they converge. However, some model chords may not intersect the object boundary and some may intersect more than once. The following rules are used to determine a unique l_i for each chord: i)for non-intersected chord, assign $l_i = l_i(B_E)$ when chord is inside an object and $l_i = l_i(B_I)$ when chord is outside an object; and ii)for multiple intersections, select l_i $\forall i$ by minimizing roughness r via dynamic programming.

EXPERIMENTS

For the first frame in a cineangiogram, segmentation is performed with Marr-Hildreth edge detector [12] which produces a Lapalacian-Gaussian sign image. The result is then unionized with a series of Lapalacian-Gaussian sign images produced from a sequence of dynamic differences between two successive frames. This operation improves the attainment of the left ventricular border due to its movement (Figure 5(a)). The model is then launched from several different initial positions and each of them results in a converged solution, but only the one with the global minimum deformity is accepted as the correct alignment (Figure 4(b)-(d)). After the final alignment is achieved, the left ventricular border of the initial frame is determined along all the chords based on the edge strength and smoothness via dynamic programming. For the second frame, the model is launched from the current matching position to align with a binary image produced by filling the interior region of the left ventricular contour detected at the current frame. After the contour determination, the process is repeated for the next frame until all the frames within a cineangiogram are processed.

This matching algorithm has been tested on cineangiograms of thirty subjects. Each of them contains 30-40 frames of a cardiac cycle with both

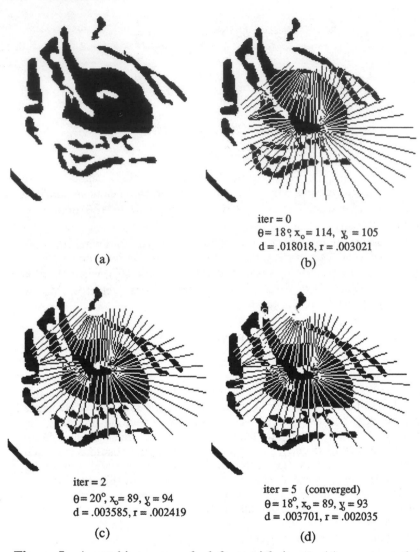

iter = 0
$\theta = 18°, x_o = 114, y_o = 105$
$d = .018018, r = .003021$

(b)

iter = 2
$\theta = 20°, x_o = 89, y_o = 94$
$d = .003585, r = .002419$

(c)

iter = 5   (converged)
$\theta = 18°, x_o = 89, y_o = 93$
$d = .003701, r = .002035$

(d)

Figure 5. A matching process for left ventricle image: (a) a segmented
binary image; (b) an initial alignment of the model;
(c) intermediate results; (d) the final match.

RAO and LAO projections. Our algorithm produced accurate left ventricle border determination for 70% of the data. The other 30% of the data failed to be analyzed due to the serious error in the inital image segmentation; the global minimum deformity is greater than $d_m = 0.01$, and the model matching failed to recognize the existence of a left ventricle. Figure 6 shows a typical example with two different frames within the same cineangiogram.

DISCUSSION

We have developed a matching algorithm utilizing a boundary constrained model to align and recognize left ventricular contours in cineangiograms. We believe that, with further improvements, this approach will give satisfactory performance of the overall processing so that all the valuable information, such as continuous volume curve, and wall motion, etc., can be reliably abtained with ease. Currently, we are exploring the direct matching of the boundary constrained model with grey scale images to eliminate the need of the image segmentation stage.

REFERENCES

1. Sun, Y.N., C.C. Li, P.R. Krishnaiah, and P.S. Reddy, "Three-Dimensional Reconstruction of Ventricle from Biplane Angiocardiograms via Equal-Division Surface", *IEEE Trans. Syst. Man. Cybern.*, Vol. SMC-19, 1989, pp. 1666-1671.
2. Van Bree, R.E., D.L. Pope, and D.L. Parker, "Improving Left Ventricular Border Recognition Using Probability Surfaces", *Personal communication*, Department of Medical Informa.ics, LDS Hospital, Univ. of Utah
3. Lilly, P., J. Jenkins, and P. Bourdillon, "Automatic Contour Definition on Left Ventriculograms by Image Evidence and a Multiple Template-Based Model", *IEEE Trans. Medical Imaging*, Vol. 8, 1989, pp. 173-185.
4. Kass, M., A. Witkin, and D. Terzopoulos, "Snake: Active Contour Models", *Proc. of the first Int. Conf. on Computer Vision*, 1987, pp. 259-268.
5. Solina, F., *Shape Recovery and Segmentation with Deformable Part Models*, PhD dissertation, Dept. of Computer and Information Science, University of Pennsyvalnia, 1987.
6. Yuille, A., D. Cohen, and P. Hallinan, "Facial Feature Extraction by Deformable Templates", Tech. report CICS-P-124, Center for Intelligent Control System, MIT, 1989.
7. Terzopoulos, D. and A. Witkin, "Deformable Models: Physically

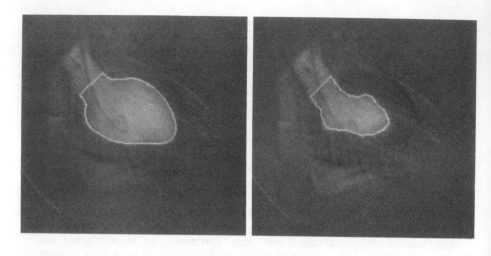

Figure 6.   A typical result of left ventricular contour recognition:
two different frames within a cineangiogram.

Based Models with Rigid and Deformable Components'', *IEEE Jour. Computer Graphics and Applications*, Vol. 8, November 1988, pp. 41-51.

8. Fan, N., *2-D Shape Recognition with Adaptive Matching of Boundary Constrained Models,* PhD dissertation, Dept. of Electrical Engineering, University of Pittsburgh, 1990.
9. Blum, H. and R. N. Nagel, ''Shape Description Using Weighted Symmetric Axis Features'', *Pattern Recognition*, Vol. 10, 1978, pp. 167-180.
10. Bookstein, F.L., ''The Line-Skeleton'', *Computer Vision, Graphics, and Image Processing*, Vol. 11, 1979, pp. 123-137.
11. Press, W.H., B.P. Flannery, S.A. Teukolsky, and W.T. Vetterling, *Numerical Recipes in C,* Cambridge Univ. Press, 1988.
12. Hildreth, E.C., ''The Detection of Intensity Changes by Computer and Biological Vision Systems'', *Computer Vision, Graphics, and Image Processing*, Vol. 22, 1983, pp. 1-27.

MEASUREMENTS OF CORONARY ANGIOGRAPHIC FEATURES

Stanley N. Hack, D.Sc.

Syracuse Research Corporation
Digital Imaging and Visualization Division
Merrill Lane
Syracuse, New York 13210

INTRODUCTION

During the past two decades, great emphasis
has been placed on evaluating coronary angiograms
by means of quantitative analyses as opposed to
the traditional qualitative and semi-quantitative
techniques. This emphasis towards quantitative
evaluation is based on the need to characterize
the efficacy of newly developed interventions
which include surgery, angioplasty, and
pharmaceutical treatments. Additionally, recent
advances in digital image processing technology
have made the quantitative analysis of coronary
angiograms a realistic goal. The primary
objective of coronary angiography is detecting
coronary arterial stenoses, or vessel narrowings,
and measuring the severity of these detected
stenoses. Hence, this paper reports on the
various techniques that have been developed to
detect and measure arterial stenoses from
angiographic images.

An angiogram is an x-ray image of a vascular
tree into which a radio-opaque contrast agent has
been injected. Without the injection of the
radio-opaque agent, the vessels are transparent to
the x-ray energy used in forming the image.
However, once radio-opaque material has been
selectively injected into the vessels of interest,
it is possible to image a two-dimensional

projection of the vessels' interiors. The cross-
sectional areas and the diameters of vessels
imaged via angiography can be easily measured from
perfect projections. The cross-sectional area is
proportional to the integral of a gray-level or
density profile perpendicular to the long axis of
the vessel in the perfect projection. The
projection diameter is the length of the
contiguous non-zero gray-levels along this density
profile. The operative phrase in this case is
"perfect projections" which, of course, is an
unrealistic expectation.

The factors that confound the notion of
"perfect projections" are as follows. First, an
angiographic projection image rarely shows only an
isolated vessel. Instead, overlapping structures
are most always superimposed on the vessel
projection image. Second, the x-ray imaging
system exhibits its own modulation transfer
function (MTF) which imposes both geometric and
densitometric distortions on the projection image.
Third, noise is always present in an angiographic
image. The noise sources include the Poisson
counting statistics of the x-ray photons, system
electronic noise, and film grain noise. Fourth,
spatial and density calibration is usually
difficult due to unknown magnifications and photon
scattering characteristics. Fifth, image blurring
is often present due to anatomical motion.
Finally, estimating the vessels' diameters from
diameter measurements made on "perfect
projections" assumes a priori knowledge of the
vessels' shapes and orientations.

Considering the great differences between
"prefect projections" and real angiographic
images, one might assume that there are a great
many difficulties in measuring vessel stenoses by
automatic techniques as well as by highly trained
humans. Such an assumption is correct. The human
inter-observer variance in assessing percent
stenosis has been reported to be as low as 8% and
as great as 37%.[1-3] As with the solutions to most
complex problems, many systems and techniques have
been proposed and implemented to automatically
measure arterial stenosis. Each technique

includes trade-offs between measurement accuracy, processing speed, and degree of required user interaction. This paper presents the salient features of many of the techniques for the automatic or semi-automatic assessment of arterial stenosis that have been reported in the literature.

STENOSIS ANALYSIS METHODS

The following steps are required to perform stenosis analysis:

1) Select image frame of interest
2) Identify vessel segment to analyze
3) Locate vessel boundaries
4) Compensate for background
5) Measure diameter and/or cross-sectional area as a function of position along the vessel segment
6) Compute stenosis ratio

The two key steps above are identifying the vessel segment to analyze (Step 2) and locating the vessel boundaries (Step 3). Several interesting approaches have been taken for identifying the vessel segment to analyze. All of these techniques require some degree of user interaction. Some of the methods used for performing this processing step are detailed in the following subsection.

The problem of locating vessel boundaries is the most complex issue of the stenosis analysis. Techniques proposed and implemented for performing this processing step fall into the following four categories:

- Manual boundary detection
- One-dimensional approach
- Two-dimensional approach
- Parametric model methods

These techniques are described in the following subsections of this report.

Vessel Segment Location

In order to detect a vessel segment's boundary and measure the stenosis ratio of a lesion within the vessel segment, it is first necessary to identify the vessel segment to be analyzed. The techniques used in this identification process range from manually tracing the vessel segment's boundaries,[4] to manually tracing the vessel segment's approximate centerline,[5-7] to specifying the approximate location of the vessel's centerline by placing multiple points along its length,[8,9] to selecting a few points (usually two) within the vessel segment.[10-12]

Fleagle *et al.*[5] require the operator to trace the vessel segment's centerline using a trackball. The traced centerline is then smoothed using a 51-point smoothing filter. Reiber *et al.*[8,9] require the operator to indicate a number of points along the approximate vessel segment midline using a sonic digitizer.

LeFree *et al.*[10] take a totally different approach. This approach presents a circle and a point indicating its center on the display screen. The user is required to drag the circle and its center until the circle's center is located inside of the vessel segment at a location midway between the two ends of the segment that the operator wishes to analyze. The operator then adjusts the circle's radius such that the vessel segment is cut at the appropriate locations proximal and distal to the circle's center. Using circular profiles, the vessel segment is tracked from the circle's center to the points of intersection of the circle and the two end-points of the vessel to be analyzed.

The author's group[11,12] requires the user to place one cursor at each end of the vessel segment to be analyzed. The system then automatically finds the path along the vascular tree that connects the two cursors.

Manual Boundary Detection

Manual boundary detection schemes provide the operator with either electronic cursors or mechanical "calipers" which must be positioned on opposite boundaries of the vessel to be measured.[13,14] The ratio of the geometric distances between the cursors positioned at a stenotic lesion and positioned at a normal location is then calculated as the stenosis ratio. This method relies upon the accuracy of the operator's placement of the cursors on the vessel boundaries, upon the operator's placement of the cursors such that a line connecting them intersects the center of the vessel at a right angle, and upon the assumption that the vessel's cross-section is round.

1-D Automatic Vessel Boundary Detection

The one-dimensional approaches to detecting the vessel boundaries are based on the analyses of gray-level profiles taken across thin sections of the vessel at right angles to the vessel's centerline. Since the basepoints of the bell-shaped profile are not clearly defined, the profile is generally searched on both sides of the centerline for the points of inflection and/or local minimums (basepoints) to be used in the determination of vessel diameter. LeFree and others[5,10] have found that using the inflection points as the positions of the vessel boundaries tends to underestimate the vessel diameter, and using the basepoints as the positions of the vessel boundaries tends to overestimate the vessel diameter. Sandor et al.[15] point out that using the full width at half maximum (FWHM) measurement for vessel boundaries is error prone since it is based on a single pixel measurement for the profile maximum. Accordingly, various investigators have used different combinations of basepoints, points of inflection, and FWHM in determining the vessel's boundaries.

The computation of the first and second derivatives of the vessel's profile is generally performed using one of two methods. The first method is discrete in that a 1-dimensional discrete convolution technique is used to determine the derivatives. The second method is pseudo-continuous in that an Nth-order polynomial is fit to the vessel profile using least squared error methods. The first and second derivatives of these polynomial functions are then computed to determine the inflection points and the basepoints.

Fleagle et al.[5] take discrete first and second derivatives of the vessel profile in order to determine the locations of the inflection points and the basepoints. They then consider the vessel boundaries to be positioned a preselected fractional distance between the locations of these two points. Once the boundary points are found for all profiles in the vessel segment of interest, the boundaries are smoothed using a directed graph method that insures that each boundary is connected but does not contain redundant neighboring points.

LeFree et al.[10] take a similar approach in that they take discrete first and second derivatives and position the boundary between these two points for each profile. Their technique for placing the boundary points is based on the location of a preselected fractional difference in gray levels between the inflection points and the basepoints. They then smooth the boundary by discarding points that create discontinuities and replacing those points by linear interpolation.

Reiber et al.[8,9] process the vessel profiles prior to locating the boundaries by subtracting an approximation of the background image from the profiles. The background estimation for each profile is determined by interpolating the gray levels of the portions of the profile well outside of the vessel over the entire profile. They then find boundary points along each background subtracted-profile via a first derivative convolution of the form [-2 -5 -5 +5 +5 +2]. They smooth the borders by repositioning the boundary points such that the boundary point on one profile is within a distance of ± 2 pixels of the boundary point on adjacent profiles. They smooth the boundaries further by fitting a second order polynomial to neighborhoods of 11 boundary points.

Spears *et al.*[6,7] fit seventh-order polynomials to each profile from which they compute the inflection points and the basepoints. They use either the inflection points or the basepoints to determine an estimate of the vessel boundary points for each profile. Then, they smooth the boundaries by fitting an Nth-order polynomial to each boundary. The centerline location is then recomputed as the midline between the two boundaries. Finally, the centerline is smoothed using a fourth-order polynomial fit.

In early reports of semi-automatically measuring vessel diameters, Selzer[16] and Beckenbach *et al.*[17] used the inflection points along each profile to determine the location of the vessel boundaries. They smoothed the detected boundary using a 21-point Fourier series. For stenosis ratio measurements, this group did not compare the diameter of the stenotic lesion with the diameter of a normal area of the vessel as most groups do. Instead, they low-pass filtered the detected boundaries by fitting a third-order polynomial to the boundaries. They then considered the new smoothed boundary to be a representation of the vessel without stenotic lesions. The stenosis ratio was then taken as the ratio of the diameters along a single profile referenced to the detected boundary and referenced to the fit polynomial.

2-D Automatic Vessel Boundary Detection

Two-dimensional techniques for locating vessel boundaries differ from the one-dimensional techniques described above in that the vessel boundaries are located using two-dimensional neighborhoods as opposed to data along individual profiles. Besson[18] uses a mathematical morphology technique that he has named a "Snake Transform" to locate vessels. Briefly, this technique searches the neighborhood of a starting pixel, known to be

within the vessel, for a pixel with minimum grey
level. (This technique assumes that pixels within
the vessel have a grey level value less than the
background. That is, vessels are dark and
background is bright.) The neighborhood of this
new pixel is then searched for a pixel with
minimum gray level value, and so on. With some
modifications to this basic transform, Besson is
able to segment vessels from the background and,
therefore, define their boundaries.

Eichel *et al.*[19] perform a global edge enhancement
on the entire image using a 2-D gradient
convolution. They then perform a sequential edge
linking (SEL) algorithm starting from a known
vessel boundary pixel. The SEL algorithm searches
for the next neighborhood pixel that is likely to
be part of the vessel boundary. A path metric
based on the path probability and its likelihood
ratio is used to determine if a traversed path is,
in fact, the vessel boundary.

The author's group[11,12] uses a neighborhood
thresholding technique to determine if a pixel is
part of the vessel. This neighborhood
thresholding technique compares each pixel to the
linear combination of its neighboring pixels.
From this comparison, a pixel is considered to be
a candidate for inclusion in the interior of a
vessel. Binary morphology is then used to locate
the vessel boundaries.

   Parametric Methods for Vessel Boundary Detection

Pappas and Lim[19] have modeled the one-dimensional
x-ray projection of a vessel profile as consisting
of components of the actual vessel profile, the
image background, the system modulation transfer
function (MTF), and additive white noise. This
model is then fit to measured profiles, and model
parameters are extracted using maximum likelihood
estimates. The vessel boundaries are then
computed from these extracted parameters.

Shmueli *et al.*[20] use a globally optimal maximum *a
posteriori* (MAP) estimate to locate a vessel

within a search window and to determine its diameter. This technique models the x-ray projection as vessel cross-section and background components. The MAP estimate locates the position and determines the diameter of a cylindrically shaped vessel within a preselected area of the x-ray projection.

DISCUSSION

As one can infer from the number of techniques that have been developed for locating and measuring blood vessels from angiograms over the past two decades, the problem is confounded by a plethora of unknowns. With so many unknowns, each technique makes use of its own set of engineering assumptions. The need to clinically measure stenotic lesions is increasing due to new interventional procedures and pharmaceuticals. Accordingly, automatic techniques for measuring stenosis ratios from angiographic images are beginning to see wide spread use in clinical environments. Unfortunately, these techniques provide estimates based on both image data and many engineering assumptions. The clinical provider must understand the limitations of the techniques used and must understand the clinical conditions under which these techniques can be used with a high degree of confidence.

REFERENCES

1. De Rouen, T.A., Murray, J.A., Owen, W., "Variability in the Analysis of Coronary Arteriograms", Circulation, 55:324-328, 1977.

2. Detre, K.M, Wright, E., Murphy, M.L., Takaro, T., "Observer Agreement in Evaluating Coronary Angiograms", Circulation, 52: 979-986, 1975.

3. Zir, L.M., Miller, S.W., Dinsmore, R.E., Gilbert, J.P., Harthorne, J.W., "Interobserver Variability in Coronary Angiography", Circulation, 53: 627-632, 1976.

4. Brown, B.G, Bolson, E., Frimer, M., Dodge, H.T., "Quantitative Coronary Arteriography: Estimation of Dimensions, Hemodynamic Resistance, and Atherma Mass of Coronary Artery Lessions Using the Arteriograms and Digital Computation", Circulation, 55:329., 1977.

5. Fleagle, S.R., Johnson, M.R., Wilbricht, C.J., Skorton, D.J., Wilson, R.F., White, C.W., Marcus, M.L., and Collins, S.M., "Automated Analysis of Coronary Arterial Morphology in Cineangiograms: Geometric and Physiologic Validation in Humans", IEEE Transactions on Medical Imaging, 8: 387-400, 1989.

6. Spears, J.R., Sandor, T., Kruger, R., Hanlon, W., Paulin, S., Minerbo, G., "Computer Reconstruction of Luminal Cross-Sectional Shape from Multiple Cineangiographic Views", IEEE Transactions on Medical Imaging, 2:49-54, 1983.

7. Spears, J.R., Sandor, T., Als, A.V., Malagold, M., Markis, J.E., Grossman, W., Serur, J.R., Paulin, S., "Computerized Image Analysis for Quantitative Measurement of Vessel Diameter from Cineangiograms", Circulation, 68: 453-461, 1983.

8. Reiber, J.H.C., Gerbrands, J.J., Booman, F., Troost, G.J., den Boer, A., Slager, C.J., Schuurbiers, J.C.H., "Objective Characterization of Coronary Obstructions from Monoplane Cineangiograms and Three-Dimensional Reconstruction of an Arterial Segment from Two Orthogonal Views", In: Applications of Computers in Medicine, pp. 93-100, IEEE Engineering in Medicine and Biology Society, New York, 1982.

9. Reiber, J.H.C., Kooijman, C.J., Slager, C.J., Gerbrands, J.J., Schuurbiers, J.C.H., den Boer, A., Wijns, W., Serruys, P.W., Hugenholtz, P.G., "Coronary Artery Dimensions from Cineangiograms - Methodology and Validation of a Computer-Assisted Analysis Procedure", IEEE Transactions on Medical Imaging, 3: 131-141, 1984.

10. LeFree, M.T., Simon, S.B., Mancini, J.G.B., Vogel, R.A., "Digital Radiographic Assessment of Coronary Arterial Geometric Diameter and Videodensitometric Cross-Sectional Area", SPIE Proceedings in Medicine XIV/PACS IV, 626: 334-341, 1986.

11. Hack, S.N., Lele, S., Streicker, M., Mostafavi, H., "Automatic Quantitative Analysis of Digital Cardiac Angiographic Sequences", SPIE Proceedings on Medical Imaging III: Image Processing, 1092: 313-325, 1989.

12. Hack, S.N., Lele, S., Streicker, M., Mostafavi, H., "Automatic Quantitative Analysis of Cardiac Angiograms", In: Proceedings of Electronic Imaging 1988, Institute for Graphic Communication, Boston, 1988.

13. Gensini, G.G., Kelly, A.E., Da Costa, B.C.B., Huntington, P.P., "Quantitative Angiography: The Measurement of Coronary Vasomobility in the Intact Anamal and Man", Chest, 60: 522-530, 1971.

14. Vanguard Instrument Corporation, XR-15 CR Coronary Analyzer, Melville, NY.

15. Sandor, T., Spears, J.R., Paulin, S., "Densitometric Determination of Changes in the Dimensions of Coronary Arteries", SPIE Proceedings on Digital Radiography, 314: 263-272, 1981.

16. Selzer, R.H., "Computer Processing of Angiograms", Symposium on Small Vessel Angiography, Glen Cove, Long Island, 1972.

17. Beckenbach, E.S., Selzer, R.H., Crawford, D.W., Brooks, S.H., Blankenhorn, D.H., "Computer Tracking and Measurement of Blood Vessel Shadows from Arteriograms", Medical Instrumentation, 8:308-310, 1974.

18. Besson, G., "Vascular Segmentation Using "Snake" Transforms and Region Growing", SPIE Proceedings on Medical Imaging III: Image Processing, 1092: 429-435, 1989.

19. Papas, T.N., Lim, J.S., "A New Method for Estimation of Coronary Artery Dimensions in Angiograms", IEEE Transactions on Acoustics, Speech, and Signal Processing, 36: 1501-1513, 1988.

20. Shmueli, K., Brody, W.R., Macovski, A., "Estimation of Blood Vessel Boundaries in X-Ray Images", Optical Engineering, 22: 110-116, 1983.

HIGH-SPEED X-RAY IMAGING IN BIOMEDICINE

Frank A. DiBianca [1], Robert J. Endorf [2], Wen-Ching Liu [2], Daniel S. Fritsch [3]

[1] Department of Biomedical Engineering, University of Tennessee, Memphis, TN; [2] Department of Physics, University of Cincinnati, Cincinnati, OH; [3] Biomedical Engineering Curriculum, University of North Carolina, Chapel Hill, NC

INTRODUCTION

This paper describes several versions of high-speed x-ray imagers which the authors have been developing over the past several years. The earlier systems [1-4] were based on field-emission (flash) x-ray tubes which provide single pulses having durations as short as 25 ns but framing rates significantly slower than 1 frame per second (fps). Most cine radiography for biomedical studies of cardiac function is carried out using conventional x-ray image intensifier systems and cameras typically allowing framing rates of 30 fps. Such rates are adequate for most natural physiological processes but are insufficient to study transient events such as bone fracture, tissue and vascular damage caused by blunt impact or penetrating projectile trauma, and similar injuries.

In our studies of flash x-ray imaging, we found that most biomedical experiments can tolerate x-ray pulse durations of from several, to several hundred, microseconds but that single-frame (i.e., non-cine) imaging is inadequate to reveal the evolution of traumatic events. Thus, we have embarked on a program to develop an x-ray imaging system (preferably all digital-electronic) capable of providing a sequence of at least 5-10 images at framing rates of 1,000 to 1,000,000 fps. We have used such a system to make images of projectiles traveling at 0.1 km/s and impacting bone, muscle and liver specimens containing simulated blood vessels and x-ray-opaque markers.

175

EXPERIMENTAL METHOD

The experiments described herein were performed on a high-speed cine x-ray system built by Hadland Photonics (Bovingdon, Herts., U.K.). The system [5] consists of an x-ray source, x-ray image intensifier, high-speed framing camera and CDD digital recording camera. The parameters of the specially-designed, demountable, microfocus x-ray source (Imax 160 MF) are summarized in Table 1. The image intensifier (Thomson CSF) has a 30 cm diameter input phosphor and blue output phosphor with submicrosecond decay time. The framing camera (Imacon 792) views dynamic events and freezes image sequences in a side-by-side image format for recording by film or a digital recording camera. Its parameters are summarized in Table 2. The output of the framing camera is recorded by a CCD (SV-553) camera which digitizes and transmits to an IBM PC/AT microcomputer an array of 1134 x 486 pixels, 12-bits deep.

An example of the spatial and temporal resolution of the camera (using light instead of x-radiation) is given in Fig. 1 which shows a 9mm caliber projectile impacting on armored cloth. The framing rate used was 100,000 fps and, for clarity, only the odd frames of an 8-frame series are shown.

Table 1. Parameters of the microfocus x-ray source.

Voltage (kV)	Current (mA)	Spot size (mm)	Exposure (ms)	Filament	Anode
40-160	20-200	0.05-0.5	0.01-1.0	demount.	demount.

Table 2. Parameters of the framing camera.

Cathode [diam]	Spectral response (nm)	Output phosphor	Framing speeds (fps)	Spatial resol. (cy/mm)	Frame on time	Framing format [# frames]
S20 UV [9cm]	180-850	P11 or P20	10,000-20,000,000	5-7	20% of cycle	[8] 16x18mm [16] 8x18mm

Fig 1.   Odd frame sequence of a 9mm calibre projectile
         impacting on armored cloth (taken at 100,000 fps).

        To demonstrate the utility of high-speed radiography for
studies of transient biomedical events, the system described
above was used to image high-speed projectiles impacting the
following biological specimens: (1) a thick bovine femur and
thinner turkey femurs; (2) bovine muscle containing x-ray opaque
markers; and (3) a section of sheep's liver containing a
simulated blood vessel filled with iodine contrast agent.

        In each case, the projectile was an 11-mm diameter steel
ball bearing fired from a 50 cm long brass tube. The steel
bearing was propelled by high pressure air and was released when
a metal pin inserted across the tube diameter was lifted by a
remotely-controlled solenoid. A light beam placed between the
end of the tube and the specimen under study was interrupted by
the passage of the ball and this initiated the x-ray exposure
and camera recording via a delay trigger of selectable delay.
Framing rates of 10,000-25,000 fps were used in these studies.

1. Bone Fracture Studies

        Several experiments were performed to study bone fracture
induced by the impact of a high-velocity (0.1 km/s) steel ball
bearing. Fig. 2 shows the impact of the ball on a thick-walled
(~3mm) bovine femur. Frames 1 and 2 show the bone before
impact. Frame 3 shows the ball at maximum penetration.
Subsequent frames show the ball recoiling from the bone at much
reduced speed. Although the bone was not fractured, a 1mm-deep
crater formed in it is seen in profile in the last five frames.

        The next image sequence (Fig. 3) shows a steel ball
fracturing a turkey femur. The initiation and extension of

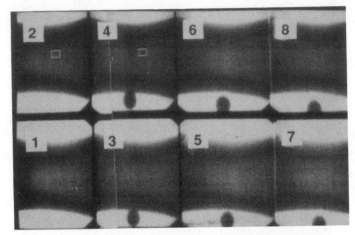

Fig 2.   Steel ball impacting bovine femur.

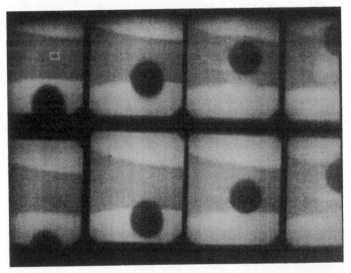

Fig 3. Steel ball penetrating turkey femur

fracture lines are seen in frames 5 and 6. In frames 7 and 8,
a section of bone has come off as evidenced by the significant
lucency in the wake of the projectile. Based on the length of
the fracture initiated between frames 4 and 5, the fracture
propagation speed is seen to exceed 0.1 km/s. Furthermore,
measurements made on the image series show that the speed of the
ball upon exiting the bone is 0.82 +- 0.02 that which it had
before impact.

    The third example of bone fracture is given in Fig. 4.
Here, another turkey femur is struck by a steel ball, but this
time the impact proceeds more explosively than in the preceding
example. Observation of the bone specimen afterwards showed
that it had indeed been blasted apart. Inspection of the last
four frames shows that the fracture fragments and edges continue
to move after the projectile has exited from the bone.

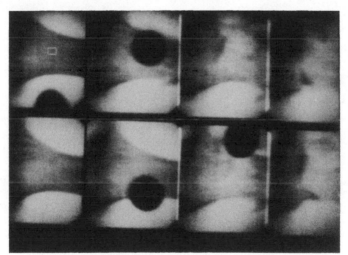

Fig 4.   Steel ball exploding turkey femur

2. Muscle Trauma

    A second series of experiments was designed to investigate
the trauma to muscle tissue caused by impact of the steel ball

at 0.1 km/s.   Since the bovine muscle sample used was quite
homogeneous, it would not have been possible to see the local
internal movement of the soft tissue directly using x-ray
imaging.   Therefore, strands of 0.75 mm solder wire were sewn
through the tissue to act as x-ray-opaque markers.   By
comparing, in Fig. 5, the positions of the wires during impact
with those of the respective wires in the first (preceding)
frame,   one   determines   the   global   (local)   horizontal
displacements of the tissue.   From the latter, one sees the
development of the shock wave in the tissue upon this type of
traumatic injury.   The shock wave is seen in Fig. 6, subject to
the limitations discussed next.

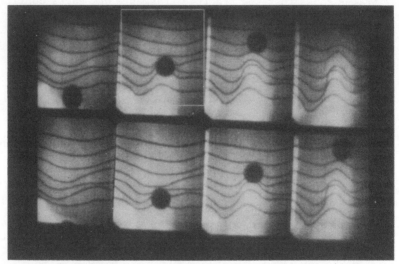

Fig 5.   Steel ball impacting bovine muscle containing wires.

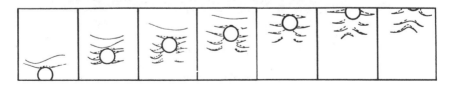

Fig 6.   Maps of shock wave propagation (displacement vectors)
         corresponding to frames in Fig 5.

This type of experiment can clearly be improved if the continuous wire markers are replaced by individual lead pellets of similar or smaller diameter. There are several reasons for this. First, the mass of the foreign material is greatly reduced and thereby its effect on the natural mechanical response of the tissue. Secondly, motion of one part of a strand is transmitted to other parts, thereby producing an erroneous component to the shock wave. Finally, one cannot determine the true three-dimensional (or even two-dimensional) motion of the tissue because one cannot observe the components of the motion along the strands, only the perpendicular components. All of these problems are eliminated or reduced by the use of individual, x-ray-opaque spheres.

3. Vascular Trauma

The final example of high-speed x-ray imaging in biomedical engineering is that of vascular trauma. This is important since much of the morbidity and mortality related to trauma is caused by vascular rupture and the attendant loss of blood.

In this study, we imbedded a simulated blood vessel into a section of sheep's liver and placed the latter in a cardboard box. The simulated blood vessel consisted of a 5mm diameter neoprene rubber tube filled with iodine contrast agent. In the first attempt to sever the artificial artery (Fig. 7), the projectile struck the vessel to one side as evidenced by the slight bending and thinning (reduced opacity) of the vessel. However, in the second attempt (Fig. 8), we achieved a direct hit as seen by the extreme distention and final severing of the simulated artery.

CONCLUSIONS

The simple experiments reported here indicate the potential usefulness of high-speed cine radiography, with frame rates in the 10,000-25,000 fps range. Not only can a host of transient processes be studied qualitatively, but quantitative data related to morphology and densitometry can also be obtained.

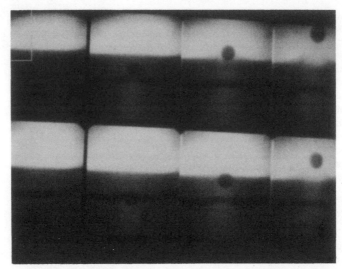

Fig 7.   Steel ball penetrating liver specimen and grazing
          imbedded simulated blood vessel containing iodine
          contrast agent.

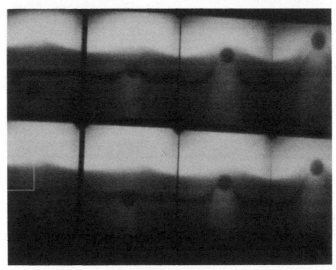

Fig 8.   Steel ball penetrating liver specimen and severing
          imbedded simulated blood vessel containing iodine
          contrast agent.

ACKNOWLEDGMENTS

    The authors are grateful to the staff of Hadland Photonics
Ltd. for their kind assistance in performing the experiments
described herein and for providing the example of optical high-
speed photography shown. In particular, we wish to thank Roger
Hadland, Frank Kosel, Brett Lawrence and Brian Speyer.

REFERENCES

1. Zoltani CK, White KJ and DiBianca FA. Progress Toward Time-
Resolved Cross Sectional Density Measurements: The BRL Computed
Tomography Facilty. Proc 8th Int Symp on Ballistics, American
Defense Preparedness Association, pp. 83-90 (IB), 1984.

2. Zoltani CK, White KJ and DiBianca FA. Multichannel Flash X-
Ray Tomography of Microsecond Events. Appl Optics, 24(23):4061-
4063, 1985.

3. Zoltani CK, White KJ and DiBianca FA. Flash X-Ray Computed
Tomography Facility for Microsecond Events. Rev Sci Instrum,
57(4):602-611, 1986.

4. Endorf RJ, DiBianca FA, Fritsch DS, Liu W-C and Burns CB.
Development of a Flash X-Ray Scanner for Stereoradiography and
CT. Proc of Int Soc for Opt Eng (SPIE), 1090:233-244, 1989.

5. Speyer BA and Hadland R. Microfocus Cine Radiography up to
1 Million Frames per Second. Proc of Int Soc for Opt Eng
(SPIE), 832:101-106, 1987.

THE DESIGN OF A MATHEMATICAL PHANTOM OF THE UPPER

HUMAN TORSO FOR USE IN 3-D SPECT IMAGING RESEARCH

JA Terry, BMW Tsui, JR Perry, JL Hendricks, GT Gullberg*.

University of North Carolina-Chapel Hill and *University. of Utah

Chapel Hill, NC and *Salt Lake City, UT.

ABSTRACT

A three-dimensional (3D) mathematical phantom has been developed for use in single photon emission computed tomography (SPECT) imaging research studies. The phantom consists of realistic models of the upper human torso and various organs which are based on the mathematical man defined in Medical Internal Radiation Dosimetry (MIRD) pamphlets 3 and 13. It includes modifications which describe the torso, lungs, left and right ventricles of the heart, liver, kidneys, rib cage and spine. The major innovations in this phantom include a torso that more accurately reflects the shape of the upper human torso and a rib cage that includes clavicles, defines the three parts of the Sternum,and reflects the shape and orientation of human ribs. This 3-D mathematical phantom has been used to describe two types of distributions: an attenuation distribution and a distribution of the uptake of Thallium-201 for cardiac SPECT imaging. By incorporating additional effects such as collimator geometry and detector response function, scatter, and noise, projection data has been simulated which resembles clinically obtained cardiac data. This simulated data is useful in the design of optimized collimators and the evaluation of reconstruction algorithms in terms of quantitative accuracy and quality of reconstructed images.

INTRODUCTION

Because of the continued prevalence of heart disease in the United States, the diagnosis of myocardial ischemia and viability remains an area of intense clinical concern. One important method used to evaluate myocardial function is single photon emission computed tomography (SPECT). Images formed with SPECT, however, are less than ideal since they reflect not only the photon source distribution, but also the attenuation effects of the body, collimator geometry and detector response

185

function, scattered photons, and statistical fluctuations in the data. Compensation for these effects is a formidable task even with the best of data. The complicated nature of image formation in SPECT, makes comparison of the many variables that might be studied (e.g. various collimator designs, attenuation or scatter) impractical or impossible in clinical settings. Physical phantoms do not realistically resemble a patient, and the image formation process is still complicated by all the above-described effects. The use of mathematical phantoms,however, to create simulated SPECT data allows for advantages over physically obtained SPECT data.

Mathematical phantoms for SPECT, which describe radiopharmaceutical uptakes in various tissues and attenuation distributions by equations or data arrays, can be used to study a variety of variables. A number of collimator designs, for instance, can be compared without the expense and time that would be necessary to physically build collimators. The presence of cardiac lesions can be modeled and included in a mathematical phantom; these lesions can be easily relocated or their sizes or intensities modified. Knowledge of the exact radiopharmaceutical distribution that is described by a mathematical phantom allows for accurate evaluation of techniques designed to improve quantitative accuracy in SPECT. While costly (in time and money) physical phantoms might also be used in the above cases, the complicating effects of attenuation, detector response, scatter, and statistical variations in the data must be present; when using mathematical phantoms these effects may or may not be included in the simulated data. In general, mathematical phantoms provide the greatest degree of flexibility in SPECT imaging research.

Specifically, a mathematical phantom has been produced that describes the organs of the upper human torso; each organ is defined by a series of equations. A radiopharmaceutical distribution of the SPECT cardiac imaging agent Thallium-201 has been produced using the above descriptions of organs. This three-dimensional source distribution has been used to create simulated two-dimensional SPECT projection data; by creating simulated projections at a number of angles, this mathematical model has been used to create SPECT data as would be produced in a clinical setting.

PHANTOM DESIGN

In choosing a mathematical phantom to produce simulated SPECT data, by far the most detailed three-dimensional model previously described is the mathematical man defined in MIRD pamphlet 3 [SYN69]. This model uses equations to describe a variety of organs and is used as a basis for the present phantom design. Unmodified, however, the original phantom is inadequate as a phantom for cardiac SPECT imaging. The modifications primarily involve the redesign of three organs.

The first modification that is required involves the heart model. In the original mathematical man phantom the heart is described as a solid, cut ellipsoid. A solid heart is too simple to reflect the cardiac wall defects that are often of most concern in cardiac SPECT imaging. This defect in the original phantom was corrected in a later MIRD pamphlet which described the walls of the heart chambers in some detail. The mathematical descriptions of the lungs (a pair of cut, half ellipsoids) and the left and right ventricles (cuts from ellipsoid shells) were obtained from MIRD pamphlet 13 [COF81].

A second modification is necessary because of the original model's description of the body torso. This torso is an elliptical cylinder (with a width:depth ratio of 2:1) that includes arm bones within this cylinder. The redefined body torso (still an elliptical cylinder) was described so as to accurately reflect the relative size, breadth and depth of an adult human male chest (a width:depth ratio of 1.46:1) [GAR71]. Additionally, arms were not included in this model since in SPECT cardiac imaging the patients' arms are often raised to the level of the head and hence are out of the field of view.

A final modification that is required involves the rib cage. The mathematical man model describes the rib cage as a set of twelve elliptical hoops perpendicular with the axis of the torso. This model is simplistic and creates unrealistic data in SPECT simulations. The increased attenuation of these "ribs" is reflected as a continuous band around the torso in some transaxial slices and not at all in other slices. In fact, portions of three to six ribs are visible in most transaxial slices of the upper human torso. The rib cage was redefined to be twelve evenly-spaced,slanted, partial hoops , a pair of clavicles (sections of a slanted hoop) and a sternum. The clavicles join the sternum, while the ribs approach but do not join the sternum. Instead, between the ribs and sternum there is a gap that is characteristically filled by cartilage in human ribs. This gap increases as one moves down the torso, with the lowest ribs displaying the largest gap. The sternum, as defined mathematically, consists of three components: the manubrium (a cut elliptical cylinder on its side),the mesosternum (a vertical elliptical cylinder) and the xiphoid process(a thin half ellipsoid).

Of the original phantom, the mathematical descriptions of three organs were retained. The kidneys (a pair of cut ellipsoids), the spine (a vertical elliptical cylinder), and the liver (a cut elliptical cylinder) were all obtained from MIRD pamphlet 3 [SNY69]. These organs were, however, relocated in the redefined upper torso. Technically, two of these organs (the liver and the kidneys) might not be considered organs of the upper

torso, however, they were included in the phantom because of their
significant uptake of Thallium-201, an imaging agent used in SPECT
cardiac studies. The placement of these organs with reference to each
other was determined with the help of various anatomical texts
[EYC11],[KOR83]. The final mathematical phantom has been defined as
a series of mathematical equations and has also been defined as a three-
dimensional data array. [Figures 1 and 2].

Figure 1. This surface rendering displays a frontal
view of the mathematical phantom.

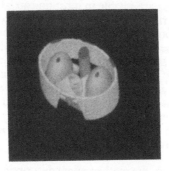

Figure 2. This surface rendering displays a vertical
view of the mathematical phantom.
Note the left and right ventricular walls.

SPECT DATA SIMULATION

Using the mathematical upper torso phantom, the three-dimensional distribution of radiopharmaceutical uptake can be described as the amount of activity per unit volume of tissue. Similarly, an attenuation coefficient distribution can be determined by taking into account that attenuation coefficients are functions of the tissue type as well as of the energy of the emitted photons from the radiopharmaceutical. With models of these two distributions completely described, simulated SPECT data, in the form of two-dimensional projections, can then be obtained.

Projections may be produced by tracing rays through the source distribution to the detector cells in the projection plane. The source distribution is integrated along a projection ray with appropriate weighting for photon attenuation. The orientation of each ray is dependent on the configuration of the collimator design which may consist of parallel-hole or converging-hole geometry. The projection rays can be modified to include the effect of detector response. The effects of scattered photons can be added into the projection data by using results from Monte Carlo simulations. Statistical variations in the data, i.e. Poisson noise, can then be added to the projections to more realistically reflect clinically obtained data . In order to reconstruct the original radiopharmaceutical distribution, two-dimensional projections must be obtained from several different views around the phantom; normally these views are arranged in a circular or half circular orbit.

Once clinically relevant projection data has been simulated, it can be used to reconstruct the 3-D source distribution. It is in comparing various reconstruction algorithms, image filters, collimator designs and imaging radiopharmaceuticals that the mathematical phantom and the simulated projection data obtained from it can prove most useful. If detection of cardiac lesions is the clinical task under consideration, observer performance in detection of such lesions can yield an evaluation of collimator performance, for example. If more accurate SPECT quantitation of source distribution is the evaluation criterion, comparisons can be easily made, between two reconstruction algorithms or two reconstruction filters, since the original distributions are known in the case of mathematical phantoms. Producing the large numbers of reconstructed images needed for observer studies or providing the certainty of a known distribution for quantitation studies are two important advantages offered by the use of mathematical phantoms.

DISCUSSION

The 3-D mathematical phantom we developed is useful in SPECT imaging research. It can be used to simulate SPECT data in evaluating various components of the imaging process. Also, it can be used to optimize collimator design without the need to build expensive collimators. Hence, by simulating the total SPECT image formation process,mathematical phantoms present realistic alternatives to expensive physical phantoms or lengthy clinical trials.

REFERENCES

COF81   Coffey JL, Cristy M, Warner, GG, Specific Absorbed Fractions for Photon Sources Uniformly Distributed in the Heart Chambers and Heart Wall of a Heterogeneous Phantom, *Journal of Nuclear Medicine/MIRD Pamphlet #13*, p 65-71, 1981.

EYC11   Eycleshymer,AC and Shoemaker,DM: *A Cross-Section Anatomy*, D Appleton-Century, New York, 1911.

GAR71   Garrett,JW and Kennedy,KW: *A Collation of Anthropometry, Volumes I and II*, Aeorspace Medical Research Laboratory March 1971.

KOR83   Koritke JG and Sick H: *Atlas of Sectional Human Anatomy, Volumes I and II*, Urban & Schwarenberg, Baltimore-Munich 1983.

SNY69   Snyder WS, Ford MR, Warner GG and Fisher HLJr, Estimates of Absorbed Fractions for Monoenergetic Photon Sources Uniformly Distributed in Various Organs of a Heterogeneous Phantom, *Nuclear Medicine/MIRD Supplement #3, Pamphlet #3*, 1969.

PSEUDO WIGNER DISTRIBUTION FOR PROCESSING SHORT-TIME BIOLOGICAL SIGNALS

C. C. Li[+], A. H. Vagnucci, MD,[+,*], T. P. Wang[+,*], and M. Sun[+]

Departments of Electrical Engineering[+] and of Medicine[*]
University of Pittsburgh
Pittsburgh, PA 15261

Abstract

Discrete pseudo Wigner distribution provides time-dependent spectral information of nonstationary time series and is a very attractive method for characterizing short-time biological signals of a limited number of samples. It can be presented as an image in the time-frequency domain, if appropriately shifted and scaled, which may show a distinctive pattern for visualization. Discriminatory features may be extracted from the Wigner distribution for use in the automatic pattern classification. This paper discusses some essential properties of the discrete pseudo Wigner distribution and shows its applications to cortisol time series for visualization and recognition of normality and abnormalities.

INTRODUCTION

Spectral analyses of stationary time series have been applied successfully to many types of bioelectric signals such as EEG, ECG, and EMG, where measurements can be obtained with relative ease for a large number of samples (1,2). Other types of biological signals such as hormonal concentration in blood plasma can only be sampled infrequently for a very limited number of samples (3-5); in such cases, it may be inappropriate to impose the assumption of stationarity. Classical Fourier analyses for such short-time, nonstationary signals give some crude spectral information at a coarse resolution. To enhance the understanding and utilization of these biological signals, the pseudo Wigner distribution can be used to provide a better estimate of the time-dependent spectral information. This paper will discuss the discrete pseudo Wigner distribution, as applied to some hormonal time series of a limited number of samples.

The principle of the Wigner distribution was first introduced by E. P. Wigner (6) in the context of quantum mechanics in 1932, and then applied to signal theory by J. Ville (7) in 1948. It did not receive much attention in the signal processing field until a decade ago; since then theories and techniques regarding the Wigner distribution have been advanced significantly, as it is applicable to obtaining time-dependent spectra of nonstationary signals (8-13). An exposition of the important mathematical background of the Wigner distribution can be found in a series of three papers by Claasen and Mecklenbrauker (8). We will briefly review some of the useful properties of the Wigner distribution in the following section. Then we will discuss its aplication to cortisol time series for visualization and classification of normal patterns versus patterns of Cushing's syndrome.

DISCRETE PSEUDO WIGNER DISTRIBUTION

Let $f(t)$ be a continuous signal of time variable t ($f(t)$ may be either real or complex), and $f^*(t)$ be a complex conjugate of $f(t)$. The Wigner distribution of $f(t)$ is defined by

$$W_f(t,\Omega) = \int_{-\infty}^{\infty} f(t+\tau/2)f^*(t-\tau/2) \ e^{-j\Omega\tau} \ d\tau \tag{1}$$

where τ is the correlation variable, $+\tau/2$ and $-\tau/2$ denote the time advance and time delay respectively. $f(t+\tau/2)f^*(t-\tau/2)$ forms a kernel function of the time variable t and the correlation variable τ. The Fourier transform of this kernel function with respect to the correlation variable τ gives the Wigner distribution $W_f(t,\Omega)$ which is a real-valued continuous function of both time t and frequency Ω.

Consider a discrete-time signal $f(nT)$ which is sampled from $f(t)$ with a sampling period T, where $nT=t$, and n is an integer. If T is equal to one time unit, the discrete-time signal can be simply denoted by a sequence $f(n)$. The discrete-time Wigner distribution is then given by

$$W(n,\Omega) = 2 \sum_{k=-\infty}^{\infty} f(n+k)f^*(n-k) \ e^{-j2k\Omega} \tag{2}$$

The discrete-time Wigner distribution $W(n,\Omega)$ is a real-valued function of the discrete variable n and the continous variable Ω; it is periodic in Ω

with its period equal to π.

In order to compute the Wigner distribution, a symmetric window function $h(k)$ of finite interval is applied to the discrete time signal $f(n)$, with its origin ($k=0$) placed at the time instant n,

$$h(k) = \begin{cases} g(k) , & -N+1 \leq k \leq N-1 \\ 0 , & otherwise \end{cases} \qquad (3)$$

where $-N+1 \leq k \leq N-1$ is the time window, and $g(k)$ can be any symmetric function, for example, $g(k)=1$. The kernel function is then computed as $h(k)h^*(-k)f(n+k)f^*(n-k)$ within the time window of length $2N-1$. If the frequency variable Ω is also discretized with $\Omega=m\Delta\Omega$ and the frequency quantization $\Delta\Omega$ is equal to $\pi/(2N-1)$, then the discrete pseudo Wigner distribution (DPWD) is given by

$$W(n,m\Delta\Omega) = 2 \sum_{k=-N+1}^{N-1} \mid g(k) \mid^2 f(n+k)f^*(n-k) \; e^{-j2km\Delta\Omega} \qquad (4)$$

$W(n,m\Delta\Omega)$ is a function of discrete-time n and discrete frequency $m\Delta\Omega$. The frequency resolution is increased by a factor of two in comparision to that of the discrete Fourier transform. In practical applications, most signals are real-valued and, with $g(k)=1$, Equation (4) can be rewritten into

$$W(n,m) = 2 \sum_{k=-N+1}^{N-1} f(n+k)f(n-k) \; e^{-j(2\pi/(2N-1))km} \qquad (5)$$

The discrete pseudo Wigner distribution has many useful properties. Here, we list six important properties; the proof and detailed discussions regarding these properties are referred to in reference (8).

1. $W(n,m\Delta\Omega)$ is real-valued.
2. $W(n,m\Delta\Omega)$ is periodic in frequency with period π, i.e.,

$$W(n,m\Delta\Omega) = W(n,m\Delta\Omega+\pi) \qquad (6)$$

This is different from the discrete time Fourier spectrum which has periodicity with period equal to 2π.
3. $W(n,m\Delta\Omega)$ has higher frequency resolution by a factor of two as compared to the discrete Fourier transform.
4. $W(n,m\Delta\Omega)$ is a bilinear transformation with respect to $f(n)$. If $\phi(n) = \sum_{k=1}^{K} f_k(n)$, then

$$W_\phi(n,m\Delta\Omega) = \sum_{k=1}^{K} W_{f_k}(n,m\Delta\Omega)$$

$$+ 2[\sum_{j > k}^{K} \sum_{k=1}^{K-1} W_{f_k f_j}(n,m\Delta\Omega)] \qquad (7)$$

5. The sum of $W(n,m\Delta\Omega)$ over its discrete frequency variable $m\Delta\Omega$ for one period is equal to the instantaneous signal power.

$$| f(n) |^2 = \frac{1}{2\pi} \sum_{m\Delta\Omega=-\pi/2}^{\pi/2} W(n,m\Delta\Omega) \qquad (8)$$

6. The sum of $W(n,m\Delta\Omega)$ over the time index n gives the energy density at the discrete frequency $m\Delta\Omega$),

$$\sum_{n=-N+1}^{N-1} W(n,m\Delta\Omega) =$$

$$| F(m\Delta\Omega) |^2 + | F(m\Delta\Omega+\pi) |^2 \qquad (9)$$

where $F(m\Delta\Omega)$ is the discrete Fourier transform of $f(n)$. This implies that if we want to evaluate the power spectrum from $W(n,m\Delta\Omega)$, $f(n)$ should be sampled with a sampling frequency larger than twice the Nyquist rate so that there will not be any aliasing problem.

CORTISOL TIME SERIES

The circadian variations of cortisol concentration in peripheral blood of both normal subjects and patients with Cushing's syndrome have been studied (3-5). In order to "synchronize" the endogenous biological clock, the experimental subjects followed the same schedule for three to five days before blood was drawn. During this adjustment period, the subjects had to rise at 7 am and go to bed at 10 pm. On the blood sampling day, the subjects maintained either a continuous recumbency or a standing position between 7 am and 7 pm, followed by recumbency from 7 pm to 7 am. Blood was sampled every half hour (T = one half hour) over a 25- to 28-hour period, providing 50 to 56 data points in each cortisol time series. Figure 1 illustrates eight cortisol time series, two in the normal category and six in the Cushing's syndrome category.

For the Cushing's syndrome category, there are three etiology classes of abnormality depending on the location of the causative tumor, which may reside in the pituitary, in the adrenal, or elsewhere (termed respectively "pituitary", "adrenal", and "ectopic" classes). It is an

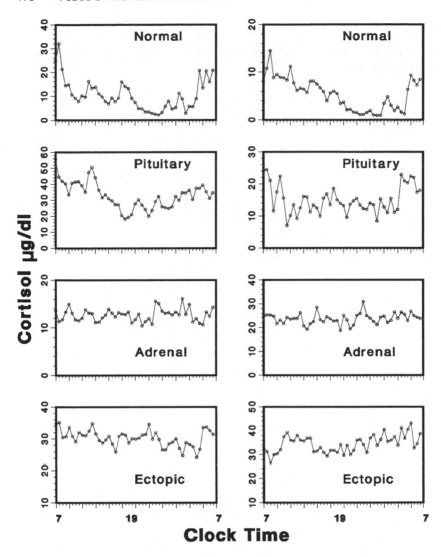

Figure 1. Eight cortisol time series of normal subjects and patients with
Cushing's syndrome (from top to bottom, two in each category:
normal, "pituitary", "adrenal", and "ectopic")

unquestionable advantage to be able to detect the abnormality and to recognize the disease class from the cortisol pattern, so as to infer the tumor location prior to confirmation with CT (or MRI) examinations and surgical exploration.

We have developed a pattern recognition system for computer classification of cortisol time series based on discriminating features extracted from Fourier analysis and Karhunen-Loeve expansion of the time series (5,14). In the present study, we applied the discrete pseudo Wigner distribution to cortisol time series to show its potential for pattern visualization and recognition of normality and abnormality.

RESULTS

We computed $W(n,m)$ directly from each cortisol time series using Equation (5). The observation window size was chosen to be 49 ($N = 25$), and the frequency quantization was $\Delta\Omega = \pi/49$. $W(n,m)$ may be presented as an image in the time-frequency domain for visual observation, if its values are all non-negative (or shifted to be so) and if appropriately scaled. Figure 2 shows eight images of $W(n,m)$, with negative values being clipped away, corresponding to the eight respective cortisol time series in Figure 1, where the horizontal axis represents the time index n ($n = 0, 1, 2, ..., 60$), the vertical axis represents the frequency $m\Delta\Omega$ ($m = -18, ..., 0, ..., 18$), and the darker region indicates a larger magnitude of $W(n,m)$. In Figure 2, from the top to the bottom, the images are the clipped $W(n,m)$ of two normal, two "pituitary", two "adrenal", and two "ectopic" cortisol patterns, respectively. It is interesting to note that such presentations show distinctive patterns for different classes under visual examination.

By observing these images, we can find that the major differences are manifested in the central portion of the time-frequency domain ($n = 13$ to 37). In that region $W(n,m)$ is most reliably computed from the summation of all 49 non-zero products of data points. It supports the observation that $W(n,m)$ has the most significant intensity information in the time interval from $n = 13$ to $n = 37$. To examine the quantitative differences among the four pattern classes, we applied Property 6 mentioned above to calculate the energy density profile along the frequency axis, utilizing only this central portion of $W(n,m)$ for each cortisol time series. The incremental energy density E_n at $m\Delta\Omega$ is denoted by

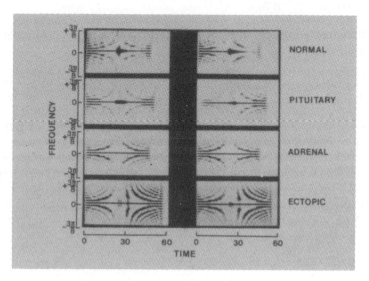

Figure 2. Wigner distribution of the eight cortisol time series in
        Figure 1,presented as images (with negative values clipped
        away) in the time-frequency domain.

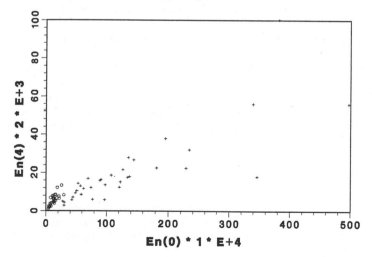

Figure 3. Distribution of normal patterns and Cushing's syndrome
        patterns in $E_n(0)$-$E_n(4)$ feature space.
        (circle : Normal; cross : Cushing's syndrome)

$$E_n(m) = \sum_{n=13}^{37} W(n,m) \qquad (10)$$

Discriminating features can be extracted from $E_n(m)$ for pattern classification.

We have processed a set of 66 cortisol time series (29 normal subject, 28 "pituitary", 2 "adrenal", and 7 "ectopic"). The normal patterns and patient patterns can be classified by using two features $E_n(0)$ and $E_n(4)$. This is shown in Figure 3 where the normal cluster and the cluster for Cushing's syndrome are apparently separated. Selected energy density values $E_n(m)$ ($m = 0, 3, 4, 7, 8$) for the cortisol time series are given in Table 1. The classification of "pituitary" versus "ectopic" can be achieved by using four spectral features $E_n(0)$, $E_n(3)$, $E_n(7)$, and $E_n(8)$. This is slightly different from the result reported in reference (15) where the auto-component of the Wigner distribution was selected and the resulting energy density features were normalized for individual cortisol time series. Work is in progress for automatic recognition of individual abnormal classes by this approach.

DISCUSSION

The discrete pseudo Wigner distribution is a very attractive method for processing short-time nonstationary biological signals. When presented as an image in the time-frequency domain, it exhibits a characteristic pattern for visualization. The extracted spectral features from the energy density profile can be effectively used for automatic pattern classification. We have illustrated its application to cortisol time series for recognition of normality and abnormality, and in the case of the latter, for differential diagnosis of disease classes. It can be advantageously applied to other kinds of biological signals for purposes of characterization and classification.

REFERENCES

1. R. G. Shiavi and J. R. Bourne, "Methods of biological signal processing," in *Handbook of Pattern Recognition and Image Processing*, eds., T. Y. Young and K. S. Fu, Academic press, Inc., NY, 1986, pp. 545-568.
2. N. V. Thakor, guest editor, Biomedical Signal Processing. *IEEE Engineerings in Medicine and Biology Magazine*, Vol. 9, No. 1, March 1990.

Table 1: Energy Density Features of Cortisol Time Series

Normal				Cushing's Syndrome				
No:	$E_n(0)$	$E_n(4)$	No:	$E_n(0)$	$E_n(3)$	$E_n(4)$	$E_n(7)$	$E_n(8)$
N 1	221195	13027	P 1	977542	25216	27132	16557	11484
N 2	161820	10926	P 2	1446050	28479	53389	19736	22274
N 3	258045	26705	P 3	1956699	10850	75702	26718	83649
N 4	127218	8802	P 4	2283947	55874	44700	82585	101010
N 5	124440	9428	P 5	1269414	45590	43136	26938	18158
N 6	109047	15476	P 6	430514	5408	11422	18270	7655
N 7	179461	13306	P 7	304301	3282	8796	9314	12965
N 8	61703	3876	P 8	300998	6067	5724	6025	9995
N 9	295149	16722	P 9	968518	7179	11865	32855	39509
N10	117952	13110	P10	774345	14307	12068	21613	13571
N11	163507	16997	P11	1359079	34393	55541	36849	20364
N12	129256	11972	P12	1210085	61372	24820	15908	32814
N13	69102	4775	P13	1377414	98314	35800	22042	51589
N14	96580	8396	P14	3822097	127354	200372	117531	182189
N15	87698	4769	P15	1343926	25557	34928	18489	18308
N16	51910	4923	P16	910600	24372	32646	19881	18960
N17	125899	14643	P17	446298	25466	14319	13899	13079
N18	74279	6235	P18	744330	28143	24469	20848	27901
N19	159230	10460	P19	892702	26226	31718	23816	20852
N20	118413	13163	P20	531961	25228	28601	14865	10633
N21	216188	15102	P21	511767	8979	20977	6732	20937
N22	136506	10746	P22	584310	28593	17026	20329	18702
N23	41614	2276	P23	624588	16822	23168	17524	23789
N24	73071	7649	P24	4973628	77516	111956	95969	88201
N25	187490	24285	P25	570898	18356	25844	14240	21810
N26	139784	7603	P26	696059	24629	33795	27494	24958
N27	77283	13736	P27	1225719	36666	30257	15346	44853
N28	148885	16878	P28	1076453	22165	37205	20954	28617
N29	49558	3598	A 1	277487	7285	9970	7890	3347
			A 2	487697	15260	18692	6211	6868
			E 1	3394271	87896	111871	34878	45896
			E 2	14196665	534662	655908	258877	370194
			E 3	8334680	184571	264436	199394	132034
			E 4	18150699	317819	398563	239083	167608
			E 5	1815392	33263	45118	27731	25984
			E 6	2338068	64267	64088	36612	27353
			E 7	3461622	160096	35968	38032	49879

3. A. H. Vagnucci,"Analysis of circadian periodicity of plasma cortisol in normal man and in Cushing's syndrome," *Am. J. Physiol.*, Vol. 236, pp. 268-281, 1979.

4. A. H. Vagnucci and D. C. C. Wang, "Analysis of periodic and random components of cortisol time series from human plasma," *J. Cybernetics and Information Science*, Vol. 2-4, pp. 73-91, 1979.

5. T. P. Wang, A. H. Vagnucci, C. C. Li, "Modeling and classification of plasma cortisol time series," in *Modeling and Control in Biomedical System*, eds., C. Cobelli and L. Mariani, Pergamon Press, Oxford, 1988, pp. 437-442.

6. E. P. Wigner, "On the quantum correction for thermodynamic equilibrium," *Phys. Rev.*, Vol. 40, pp. 749-759, 1932.

7. J. Ville, "Theorie et applications de la notion de signal analytique," *Cables et Transmission*, 2^eA, 1, 61-74, 1948.

8. T. A. C. M. Claasen & W. F. G. Mecklenbrauker, "The Wigner distribution - a tool for time-frequency signal analysis, Part I, Part II, Part III," *Philips J. Res.*, Vol 35. pp. 217-520, 276-300, 372-389, 1980.

9. F. Boudreaux-Bartels, *Time-frequency Signal Processing Algorithms: Analysis and Synthesis Using Wigner Distributions*, Ph.D Dissertation, Rice University, 1984.

10. W. Martin & P. Flandrin, "Wigner-Ville spectrum analysis of nonstationary processes," *IEEE, Trans. Acoust., Speech, Signal Processing*, Vol. ASSP-33. pp. 1461-1470, Dec. 1985.

11. J. C. Andrieux, M. R. Feix, G. Mourgues, P. Bertrand, B. Lzrar, and V. T. Nguyen, "Optimum smoothing of the Wigner-Ville distribution," *IEEE Trans. Acoust., Speech, Signal Processing*, Vol. ASSP-35, pp. 764-769, 1987.

12. M. Sun, *The Discrete Pseudo Wigner Distribution: Efficient Computation and Cross-Component Elimination*, Ph. D. Dissertation, Univer. of Pittsburgh, 1989.

13. M. Sun, C. C. Li, L. N. Sekhar, and R. J. Sclabassi, "Efficient computation of discrete pseudo Wigner discribution," *IEEE Trans. Acoust., Speech, Signal Processing*, Vol. ASSP-37, pp. 1735-1742, 1989.

14. A. H. Vagnucci, T. P. Wang, V. Pratt, C. C. Li, "Classification of plasma cortisol patterns in normal subjects and in Cushing's syndrome," *IEEE Trans. on Biomedical Engineering*, (to appear)

15. T. P. Wang, M. Sun, C. C. Li and A. H. Vagnucci, "Classification of abnormal cortisol patterns by features from Wigner spectra," *Proc. of International Conference on Pattern Recognition*, Atlantic City, NJ, June, 1990.

Electrical Impedance Tomography in Three Dimensions

J.C. Goble, D.G. Gisser, D. Isaacson and J.C. Newell

Rensselaer Polytechnic Institute
Troy, NY 12180

Introduction

The inverse problem in Electrical Impedance Tomography (EIT) considers the reconstruction of the resistivity distribution inside an object from electrical measurements made on the periphery. Although superficially similar, this task is fundamentally different from that encountered in X-ray Computed Tomography or Positron Emission Tomography, where the photon paths through the medium are essentially straight lines. In contrast, the current paths in EIT are functions of an unknown resistivity distribution, giving rise to a non–linear reconstruction problem.

In this paper, we briefly review progress in the three dimensional EIT problem. Specifically, we

- describe the physiological basis for EIT,

- introduce notation for the 3–D problem,

- demonstrate results from a simple reconstructor that assumes a uniform resistivity distribution along the longitudinal dimension of the tank.

The Physiological Basis for EIT

Biological tissues contain free charge carriers that permit them to act as (relatively poor) electrical conductors. This ability to conduct varies substantially among various types of tissue— some typical values of resistivity for tissues of interest have been tabulated by Baker [1] and are presented as Table 1. The goal of EIT is to compute and display the spatial distribution of resistivity inside the body.

Also present in tissue are bound charges that result in dielectric properties. In the presence of an electromagnetic field, these charges give rise to displacement currents that are manifested as phase shifts between voltage and current. Since this permittivity is also spatially variable, we

Material	Resistivity (ρ) ohm-cm
Blood	150
Plasma	63
Cerebrospinal fluid	65
Urine	30
Skeletal muscle	300
Cardiac muscle	750
Lung	1275
Fat	2500
Copper	1.724×10^{-6}

Table 1: Typical Tissue Resistivity Values

can also strive to produce an image of the dielectric constant. Isaacson and Fuks [4] have shown that the real and imaginary components of the complex tissue impedance are separable, at least to first order. We restrict the present discussion to the resistive impedance for sake of brevity.

Although the research at Rensselaer focuses on biomedical imaging, there are many other areas of application. The technique may find utility in geophysical exploration, non– destructive testing, environmental monitoring and other fields where physical changes result in a spatially variable distribution of resistivity.

Some Underlying Concepts

In EIT, we apply small alternating currents through electrodes attached to the skin, and measure the resulting voltages. From Maxwell's equations, we can obtain the governing equation,

$$\nabla \cdot \sigma \nabla u = 0 \qquad (1)$$

inside the body Ω, where u is the electric potential and σ is the conductivity. Since currents are applied through electrodes on the skin, a current density is established whose inward pointing normal j is

$$\sigma \frac{\partial u}{\partial \nu} = j \text{ on } \partial \Omega. \qquad (2)$$

We now introduce a notation for the three dimensional EIT problem.

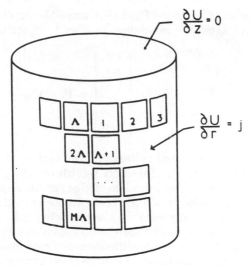

Figure 1: Tank and Electrode Geometry

Notation

Consider a right cylindrical volume with a three-dimensional electrode array on the surface as depicted in Figure 1. Let M denote the number of electrode rings and μ the ring index. Similarly, let Λ denote the number of electrodes on any ring and λ the electrode index. Consequently, the total number of electrodes is $L = M\Lambda$.

Using external electronics, steady state patterns of current are established on the electrode array, and as will be seen, measurements resulting from $L - 1$ orthogonal current patterns over the L electrodes are used to reconstruct the resistivity distribution within the object. Denote the kth current pattern by the L–vector I^k and the corresponding voltage measurement by another L-vector V^k, so that

$$I^k = \begin{bmatrix} I_1^k \\ I_2^k \\ I_3^k \\ \vdots \\ I_L^k \end{bmatrix} ; \qquad V^k = \begin{bmatrix} V_1^k \\ V_2^k \\ V_3^k \\ \vdots \\ V_L^k \end{bmatrix} . \tag{3}$$

We constrain each of the applied current patterns so that $\sum_{l=1}^{L} I_l = 0$, and to preserve this symmetry, we renormalize the measured voltages so that $\sum_{l=1}^{L} V_l^k = 0$.

We will refer to an operator $\mathbf{R}(\rho)$ that maps the applied currents into the measured voltages. We can define another L-vector U^k

$$U^k = U(\rho) = \begin{bmatrix} U_1^k \\ U_2^k \\ U_3^k \\ \vdots \\ U_L^k \end{bmatrix} = \mathbf{R}(\rho)I^k \tag{4}$$

which represents the predicted voltages on the surface of the tank due to some resistivity guess ρ. The **forward problem** consists of predicting the voltages on L electrodes from knowledge (or an assumption) about the resistivity distribution inside the tank. In the case of a homogeneous cylinder, we can analytically solve for the predicted voltages U^k using increasingly more complete models for the boundary conditions. The more general case requires use of finite difference or finite element techniques to solve the forward problem for an arbitrary resistivity distribution.

Because $\mathbf{R}(\rho)$ is self–adjoint, it has at most

$$N \leq N_{max} = \frac{L(L-1)}{2} \tag{5}$$

degrees of freedom. Consequently, we can subdivide the volume of interest into at most N_{max} voxels, or volume elements, on which the resistivity is considered piece–wise constant. Hence we will compute a resistivity vector of length at most N such that

$$\rho(p) = \begin{bmatrix} \rho_1(p) \\ \rho_2(p) \\ \rho_3(p) \\ \vdots \\ \rho_L(p) \end{bmatrix} = \sum_{n=1}^{N} \rho_n \chi_n(p) \tag{6}$$

and

$$\chi_n(p) = \begin{cases} 1 & \text{if } p \in \text{voxel n} \\ 0 & \text{otherwise.} \end{cases} \tag{7}$$

Figure 2 depicts the geometry of a typical voxel as used in our existing reconstructor. The **inverse problem** consists of predicting $\rho(p)$ throughout the body from knowledge of the applied currents and measured voltages.

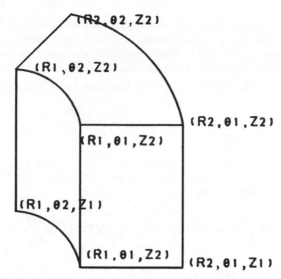

Figure 2: Voxel Geometry and Notation

Current Research Efforts

Although solving the inverse problem in EIT is an attractive goal, a variety of researchers have attacked the problem with little success. Yorkey [14] reviewed the performance of six different two dimensional EIT algorithms and concluded that a modified Newton–Raphson method provided best results on simulated data. Other techniques compared included perturbation techniques [9, 12], equi–potential back–projection techniques reminiscent of CT algorithms [3, 2] and double–constraint techniques [13].

Simske [11] has implemented a simple two dimensional algorithm based on the Newton–Raphson model. The results described in the next section represent an extension of these concepts to three dimensions. This author [6] recently demonstrated images from experimental data using three dimensional electrode arrays. Efforts of others have focused on correcting existing two dimensional techniques for out–of–plane current flow [8, 7].

Results

We have constructed a three–dimensional saline tank as in Figure 1. with $\Lambda = 16$ and $M = 4$. The tank radius and height are 15.0 cm and 25.0 cm respectively. The electrodes were constructed of polished 304 stainless steel with an area of 25.8 cm^2. Conductivity of the saline filling the tank was adjusted to be approximately 750 Ω–cm. Variously resistive and conductive targets were introduced into the tank to evaluate the performance of reconstruction algorithms.

The goal is to minimize, in the least square sense, an error function $E(\rho)$,

$$E(\rho) = \sum_{k=1}^{L-1} ||V^k - U^k||^2 = \sum_{k=1}^{L-1}\sum_{l=1}^{L}(V_l^k - U_l^k)^2. \tag{8}$$

Again, V^k represents the voltages experimentally measured for the kth current pattern and U^k represents the predicted voltage for that pattern produced by the forward solver. At a minimum,

$$0 = \frac{\partial E}{\partial \rho_n} = F(\rho) = -2\sum_{k=1}^{L-1}\sum_{l=1}^{L}(V_l^k - U_l^k)\frac{\partial U_l^k}{\partial \rho_n}. \tag{9}$$

We then can compute a new estimate,

$$\rho^{i+1} = \rho^i - [F'(\rho^i)]^{-1}F(\rho^i), \tag{10}$$

where

$$F' = \frac{\partial^2 E}{\partial \rho_i \partial \rho_l} = \begin{bmatrix} \frac{\partial F_1}{\partial \rho_1} & \cdots & \frac{\partial F_1}{\partial \rho_N} \\ \vdots & \ddots & \vdots \\ \frac{\partial F_N}{\partial \rho_1} & \cdots & \frac{\partial F_N}{\partial \rho_N} \end{bmatrix}. \tag{11}$$

We have implemented a náive reconstructor called **N2.5D** that assumes $\partial U/\partial z = 0$ between rows of electrodes as well as on the ends of the bounding cylinder. Each image "slice" is comprised of resistivity values computed to be piece–wise constant on a set of 120 voxels, or "cylindrical cuboids" for each row of electrodes.

A 3.5cm tall cylindrical resistive target of total volume 45 cm^3 was introduced into the test tank at the level of the third row of electrodes. The target was radially located halfway between the electrodes and the center of the tank. The saline in the tank was adjusted to have a resistivity of approximately 1500 $\Omega - cm$ and the target resistivity was effectively infinite. The Rensselaer **ACT-2** [4, 5, 10] instrument was used to apply current and measure the resulting voltages.

A set of fifteen orthogonal current patterns were established on the electrode array, and resulting voltages were recorded. These patterns were the "stacked cosine" [6] patterns that diagonalize the resistivity operator in two dimensions.

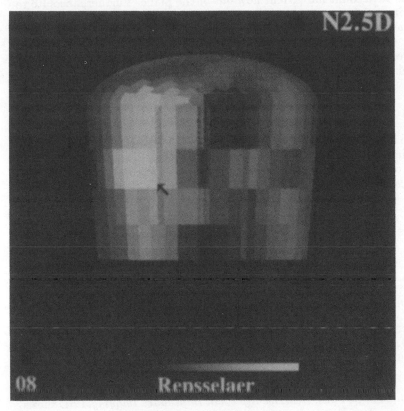

Figure 3: **N2.5D** Reconstruction

N2.5D was used to reconstruct the internal resistivity of the tank. A view through the center of the target is shown in Figure 3. Although the target was correctly localized radially, it appears elongated due to failure to account for current flow out of the plane. Further, absolute resistivity measurements are inaccurate due to the underestimated volume of conduction through the tank. **N2.5D** produced a maximum resistivity contrast of 12.2 $\Omega - cm$. It is expected that both spatial contrast and resistivity contrast will be substantially improved by more accurately modeling current flow in the tank.

Conclusion

Three dimensional reconstruction algorithms for EIT represent a substantial computational challenge. This work has promise of providing new tools for the solution of this inverse problem that may have wide applicability.

References

[1] Lee E. Baker. Principles of the impedance technique. *IEEE Engineering in Medicine and Biology*, 3(5):11–15, 1989.

[2] D. C. Barber and B. H. Brown. Applied potential tomography. *Journal of Physics E: Scientific Instrumentation*, 17:723–733, 1984.

[3] D. C. Barber, B. H. Brown, and I. L. Freeston. Imaging spatial distributions of resistivity using applied potential tomography. *Electronic Letters*, 19:933–935, 1983.

[4] Luiz Felipe Fuks. *Reactive Effects in Impedance Imaging*. PhD thesis, Rensselaer Polytechnic Institute, Troy, NY, 1989.

[5] D.G. Gisser, D. Isaacson, and J.C. Newell. Current topics in impedance imaging. *Clinical Physics and Physiological Measurement*, 8(Supplement A):39–46, 1987.

[6] John C. Goble and David Isaacson. Optimal current patterns for three dimensional electric impedance tomography. *Proceedings, IEEE Engineering in Medicine and Biology*, 2:463–465, 1989.

[7] Y. Ziya Ider and Nevzat G. Gencer. An algorithm for compensating for 3d effects in electrical impedance tomography. *Proceedings, IEEE Engineering in Medicine and Biology*, 2:465–466, 1989.

[8] Y. Ziya Ider, Nevzat G. Gencer, Ergin Atalar, and Haluk Tosun. Electrical impedance tomography of translationally uniform cylindrical objects with general cross–sectional boundaries. *IEEE Transactions on Medical Imaging*, 9(1):49–59, 1990.

[9] Y. Kim, J.G. Webster, and W.J. Tompkins. Electrical impedance imaging of the thorax. *Journal of Microwave Power*, 18:245–257, 1983.

[10] J. C. Newell, David G. Gisser, and David Isaacson. An electric current tomograph. *IEEE Transactions on Biomedical Engineering*, 35(10):828–833, 1988.

[11] Steven J. Simske. An adaptive current determination and a one-step reconstruction technique for a current tomography system. Master's thesis, Rensselaer Polytechnic Institute, Troy, NY, 1987.

[12] L. Tarassenko and P. Rolfe. Imaging spatial distributions of resistivity– an alternative approach. *Electronic Letters*, 20:574–576, 1984.

[13] Alvin Wexler, B. Fry, and M. R. Neuman. Impedance-computed tomography algorithm and system. *Applied Optics*, 24:3985–3992, 1985.

[14] Thomas J. Yorkey. *Comparing Reconstruction Methods for Electrical Impedance Tomography*. PhD thesis, University of Wisconsin - Madison, Madison, WI, 1986.

ULTRASONIC TECHNIQUE AND ARTIFICIAL INTELLIGENCE:

DIFFERENTIATION OF TISSUE TYPES

Pushkin Kachroo, T. A. Krouskop*, Anjala Kachroo,
J. B. Cheatham, P. Barry*
Rice University and Baylor College of Medicine*

4127 Long Grove Dr., Seabrook, Texas 77586

INTRODUCTION

Pressure sores have been and remain a serious problem
in the management and rehabilitation of the chronically ill patient
(Nola). These sores are a common and serious problem facing
all bedridden and wheelchair confined patients. The primary
cause of pressure sores is unrelieved pressure, shearing force, or
both, leading to ischemia of localized areas of skin and
subcutaneous tissue, usually overlying weight-bearing bony
prominences (sacrum, ischium, greater torchanter) (Reichel)
(Nola). Experimental studies have not only shown that pressure
is the main factor, but also that there exists an inverse relationship
between the amount of pressure applied and the length of time of
application in the production of pressure sores (Lindan). The
incidence of pressure sores among paraplegic and quadriplegic
patients ranges from 25 to 85%, depending on the medical and
nursing care received (Dinsdale). An estimated 7 to 8% of deaths
in this group can be attributed directly to pressure sores (Freed,
et al). Prolongation of hospitalization, amount of suffering and
delay of rehabilitation are definitely mentionable on account of
the sores. Not only hospitalized patients are at risk, even persons
who are simply inactive and have other medical problems such
as poor blood circulation or incontinence are at risk of developing
ulcers. If the degree of pressure applied to a tissue is high
enough to interfere with tissue circulation and if the duration is

211

also long enough, it causes skin breakdown (Lindan). Thus
people who already have bad circulation are at higher risk of
getting pressure sores. Obviously pressure sores have bad
effects economically. In a study done for the year 1966, it was
discovered that each ulcer increased the cost of medical care by
$5000 (Schell, et al). In a study done for the year 1969, it was
noted that insurance companies allotted 25% of the anticipated
expenses of a spinal cord injury for the treatment of pressure
sores (Griffith). A study done for the year 1979 estimated the
expense created by a single sore close to $14,000 (Vistines).
Contemporary prices for medical care would obviously make the
dollar figures considerably higher. Early identification of tissue
damage is of considerable importance. In many cases, early
clinical intervention can prevent breakdown (Satsue Hagisawa, et
al). Early detection of tissue distress is routinely practiced by
conducting regular inspection of skin color at sites over bony
prominences, but these observations have limited significance.
The observation of skin color, for the early detection of the
formation of pressure sores, is non-quantitative and non-specific.
Research has been done using different approaches in an
attempt to detect the muscle damage early. Hagisawa, et al,
conducted a study which was undertaken to seek biochemical
indicators in blood which would indicate the onset of a pressure
sore. Their work showed the potential of serum creative
phosphokinase (CPK) as a systematic indicator of the sore
development. Another project (Bennett, et al) was carried out in
order to study the skin blood flow in seated geriatric patients.
Dowd, et al worked on skin viability measurement using the
transcutaneous oxygen monitor. Thermographic study by Lewis,
et al used temperature of the skin as a criterion. Ultrasound has
been used in some studies also. Bhagat, et al did work which
dealt with attenuation and back scattering in freshly excised
animal tissues. One in vivo ultrasound experiment done by Chu,
et al also used attenuation and the back scattering property of the
tissue.

METHODOLOGY

Thus far, there have been no studies to determine the
relationship between the health and the mechanical properties of
human tissue. Although tissue breakdown can occur for many

reasons, pressure bearing areas are most likely to develop ulcers. The most common reason for pressure sores is a lack of blood supply to the tissue, causing the tissue to die and stiffen (high elastic modulus). Ultrasound technology is used in the present work to differentiate between tissue types and then highlight the region of a possible development of a pressure sore.

Objectives

(1) To modify the existing ultrasound technology in order to measure the stiffness of the various tissue types.
(2) To develop a theoretical basis for pressure sore identification.
(3) To pave the way for future work in the same domain.

Apparatus

The ultrasound system developed for this study has the following main components.

(1) tissue vibrator (perturbation) having three ultrasonic transducers embedded in it
(2) perturbator box to control vibrations of a specific frequency for the tissue vibrator
(3) computer, so as to analyze and collect a set of data
(4) Doppler box, in order to communicate between the perturbator and the computer

Procedure

The perturbator creates vibrations on the surface of the skin. The transducer finds the amplitude of the vibrations at different depths, which is then plotted on the computer screen. The software then calculates the elastic moduli, differentiates between different tissue types, and finds out if there is a development towards a pressure sore formation and if so, at which depth.

MATHEMATICAL DEVELOPMENT

Pertaining to the elastic properties of materials, the biological material beneath the outermost layer of the skin can be broadly divided into linear visco-elastic material (soft tissue and muscle), linear elastic solid (bone), non-linear elastic solid (blood vessel wall) and nonlinear fluid (blood). Here we are basically concerned with propagation of waves through these different types of materials.

The equation for an acoustic (irrotational) wave propagated along the x_1 axis is given by,

$$\rho\frac{\partial^2 X_1}{\partial t^2} = (K + \frac{4}{3}G)\frac{\partial^2 X_1}{\partial x_1^2}$$

The solution to this equation for a visco-elastic fluid can be assumed to be of the form

$X_1 = X_p\cos(wt) + Y_p\sin(wt)$

where X_F and Y_F are time independent. Ignoring Y_F

$$E = \frac{-3\rho w^2 A}{2\frac{\partial^2 A}{\partial x_1^2}}$$

Due to absorption in the viscous media, the amplitude of the wave decays exponentially with X_1. So A can be written as

$$A = \alpha e^{-\gamma x}$$

where α and γ are constants and x denotes the direction X_1. Best squares fit techniques can be used to calculate α and γ. Now we have

$$E = -\frac{3\rho w^2}{2\gamma^2}$$

Conditions near the boundary have to be analyzed. The reflected wave travels back to the first medium and the amplitude of the reflected wave is related to the incident wave as

$$A_r = \frac{R_1 - R_2}{R_1 + R_2}$$

where $R_1 = \rho_1 c_1$, $R_2 = \rho_2 c_2$, ρ = density of each material, c = velocity, and A_r = ratio between reflected and incident amplitudes

Consider the case when the reflected wave dies out passing through a single constant elastic modulus material, as shown in the following figure.

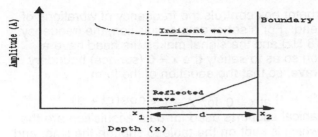

Figure 1   Near Boundary Condition

The plot of A versus x should look like

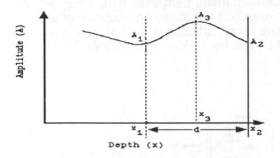

Figure 2 Boundary Amplitude Crest

Sometimes multiple reflections can also occur. A thin medium like blood vessel wall also creates a bump on the amplitude versus depth curve but to a very small degree.

INSTRUMENTATION

The instrumentation for this project broadly consists of an ultrasound head, a perturbator box, mechanical supports, a computer, and a doppler displacement instrument.

The ultrasound head looks like a circle in plan view and an inverted T in the side elevation. The perturbator has the transducer embedded in it. Thus, it has a dual purpose, one as a vibrator head and the other as a transducer.

216   P. KACHROO ET AL.

The perturbator box controls the frequency of vibrations of the perturbator head. For a set of data acquisition, the frequency is held constant (8 Hz) and the signal makes the head have a sinusoidal vibration so as to satisfy the x = 0 (surface) boundary condition of the wave, so that the equation of the form

$$Ae^{-\beta x} \cos(ct - dx + \emptyset) \text{ at } x = 0 \text{ for } A = 1 \text{ is } \cos(ct + \emptyset).$$

The mechanical supports used for data acquisition are the table, a cushion which is kept on the table to support the limb, and the hydraulic jack which has a pedal, with the help of which the height of the perturbator head can be adjusted according to need.

An Apple IIe computer has been used for data acquisition. For analysis, a Sun Workstation has also been used.

The doppler displacement instrument uses a 10 MHz crystal to generate the waves and uses delay and doppler effect to calculate the flow of tissue for a particular instant of time at a particular depth. Figure 3 shows the overall assembly of the instrumentation.

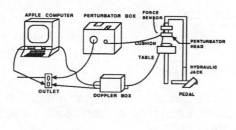

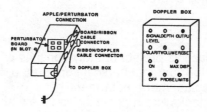

Figure 3 Perturbator Equipment Assembly

STATISTICAL ANALYSIS

Data was collected on different regions of the arm on six subjects on various days. The data was in the form of files containing depth values and the corresponding amplitude of vibrations. The amplitude values having major oscillations which indicate bone affected regions were eliminated. Each data set was divided into parts for different depth ranges. For each range, best fitting was done to get the fit parameters and then elastic modulus of each region was calculated. On the basis of tissue differentiation performed manually, the modulus values were kept in separate sets for skin, fat, and muscle. The effective modulus for blood vessel area was also calculated. Statistical analysis was done on each set of data. This statistical analysis of the data on skin, fat, muscle, and blood is shown in the tables which follow. Although these values are not very consistent as can be seen on the plots as well as the statistical outputs, and also by noting the changes in the output, by removing three data values for muscle, the tissue types seem to fall in the following general ranges.

modulus for muscle	0.1 to 5 psi
modulus for fat	10 to 200 psi
modulus for skin	3000 and above
effective modulus for blood vessel region	0.2 psi and less

These values combined with the mathematical analysis developed to understand the amplitude versus depth patterns were used to formulate the software algorithms.

SOFTWARE DEVELOPMENT

Figure 4 shows how all the computer programs are connected together. PERT.DATA/PERT.BIN use the interrupt clock to average the tissue flow and find out the maximum flow, i.e. the amplitude for various depths. PERT.COMB is used for combining the different data files. The analysis program is written in C language which calls functions fat() to eliminate the fat region, fit() to calculate the best fit parameters, modulus() to calculate the elastic modulus of the region, bone() to eliminate the bone affected region, and blood() to eliminate the region affected by blood. The rest of the region is processed by the program muscle.c to look for possible pressure sore development on the basis of flat curves or in other words, high stiffness regions.

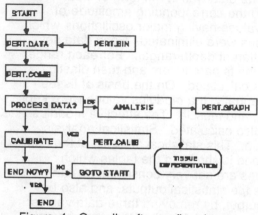

Figure 4   Overall software flowchart

CONCLUSIONS

Since at this point extensive data on patients with pressure sores has not been taken, graphical outputs for different stiffnesses were studied which can be used for further research in developing and validating the final computer package.   The software which can differentiate tissues, needed ultimately for the development of a complete software package which can identify pressure sores in their early stages, is complete. The expected patterns created on the amplitude versus depth curve by the varying degrees of pressure sores, were studied graphically. This can prove to be useful in the development of the final software for pressure sore identification.

The aim of this project was to be able to detect pressure sores at early stages. To achieve that, data on patients with varied degrees of pressure sores needs to be taken. Next, an exact algorithm in the form of a computer program needs to be evolved from the study done on that data, which will detect the region where the pressure sore is building up. This program should also determine the intensity of the formation of the pressure sore.

NOMENCLATURE

x_1, x_2, x_3	Rectangular coordinates
X_1, X_2, X_3	Displacement in direction of x_1, x_2, x_3 axes
E	modulus of elasticity in tension
G	modulus of elasticity in shear
ρ	density
ω	angular frequency
A	amplitude

ACKNOWLEDGEMENTS

The authors would like to express gratitude to the Mobility Foundation for sponsoring this research project.

REFERENCES

Bennett, L., D. Kavner, B. Y. Lee, F. S. Trainor, and J. M. Lewis
    1981 "Skin blood flow in seated geriatric patients." Arch
    Phys Med
    Rehabil 62: 392-98.
Bhagat, P. K., V. C. Wu
    1980 "Attenuation and backscattering in freshly excised
    animal tissues."
    IEEE Trans Biomed J 27(2): 76-83.
Chu, W. K., Anderson, Imray, Batros, and Cheung
    1985 "A Microcomputer-based in-vivo ultrasound tissue
    differentiation system." J of Clinical Engineering. 10(4).
Dinsdale, S. M.
    1974 "Decubitus ulcers: Role of pressure and friction in
    causation." Arch Phys Med Rehabil 55: 147.
Dowd, G. S., K. Linge, and G. Bentley
    1985 "Skin viability measurement using the
    transcutaneous oxygen monitor." In Whittle, M. and D.
Freed, M. M., H. J. Bakst, and D. C. Barrie
    1966 "Lefe expectancy, survival rates, and causes of
    death in civilian patients with spinal cord trauma." Arch
    Phys Med Rehabil 47: 457.

Gaeton, T. Nola and Lavs M. Vistines
     1980 "Differential response of skin and muscle in the
     experimental production of pressure sores." Plastic and
     Reconstructive Surgery.   66(5).
Griffith, B. H.
     1963 "Advances in the treatment of decubitus ulcers."
     Surg Clin North Am 43: 245.
Hagisawa, Satsue, Ferguson-Pell, V. Palmieri, and G. Cochran
     1988 "Pressure Sores: A Biochemical Test for Early
     Detection of Tissue Damage." Arch Phys Med Rehabil 69.
Krouskop, T. A., D. R. Dougherty, and F. S. Vinson
     1987 "A Pulsed Doppler Ultrasonic System for making
     Noninvasive Measurements of the Mechanical Properties
     of Soft Tissue." Journal of Rehabilitation Research. 24(2):
     1-8.
Lewis, D. W., and R. E. McLaughlin
     1971 "Thermographic studies of skin subjected to
     localized pressure."
     Am J Roentgenol Nuc Med 113: 749-54,
Lindan, O.
     1961 "Etiology of decubitus ulcers: An experimental
     study." Arch Phys Med Rehabil 42: 774.
Reichel, S. M.
     1948 "Shearing force as a factor in decubitus ulcers in
     paraplegics." J A M A 166: 762.
Schnell, V. C. and L. E. Wolctoo
     1966 "The etiology, prevention, and management of
     decubitus ulcers."Mo Med 63: 109.
Vistines, L. M.
     1979 "Pressure Sores: Etiology and treatment: National
     Merit Research
     Proposal." VA Central Office.

ACQUISITION AND IMAGE RECONSTRUCTION METHODS FOR IMPROVED CARDIAC SPECT IMAGING

BMW Tsui, XD Zhao, EC Frey, JR Perry and *GT Gullberg.

Dept. of Radiology and Curriculum in Biomedical Engineering, Univ. of North Carolina at Chapel Hill, Chapel Hill, NC. *Univ. of Utah, Salt Lake City, UT.

ABSTRACT

Nuclear medicine is a major imaging technique for diagnosis of cardiac diseases. Nuclear ventriculogram and planar Tl-201 imaging are particularly useful in the study of regional cardiac motion and myocardium perfusion defects, respectively. With the advance of single-photon emission computed tomography (SPECT), we are able to obtain images with improved contrast and detail which was impossible before. However, SPECT imaging technique is affected by statistical noise fluctuations, and physical factors such as finite detector spatial resolution, and photon attenuation and scatter in the patient. These result in degradation in image quality and quantitative information which are important in clinical diagnosis. We will report new acquisition and image reconstruction methods which provide reduction of image noise and compensation for the physical factors that affect SPECT. Simulation and clinical images will be presented to demonstrate the efficacy of the improved acquisition and image reconstruction methods in cardiac SPECT imaging.

INTRODUCTION

Cardiac diseases are among the most frequent causes of death in this country today. The early detection and diagnosis of myocardial dysfunction are major goals in health care. Nuclear medicine is particularly useful due to its unique capabilities in studying regional cardiac motion as well as functional perfusion of the myocardium. Such information is difficult if not impossible to obtain with other means.

Single-photon computed emission tomography (SPECT) is an imaging method which combines conventional nuclear medicine imaging with computed tomography techniques. It provides three-dimensional reconstructed images which are superior to conventional nuclear medicine images in terms of image contrast and detail.

However, the quality of SPECT images is affected by the detector system which determines the statistical noise and spatial resolution in the reconstructed image. Photon attenuation and scatter in the patient severely degrade the quantitative information and contrast of the image.(*1, 2*) In order to improve the quality and accuracy of quantitative information of SPECT images, new detector system designs and image processing and reconstruction methods have been devised. In this paper, we describe some significant progress in these areas.

NEW COLLIMATOR DESIGNS

Most SPECT imaging systems are based on single or multiple radiation detectors mounted on a stationary or rotating gantry. In commercial system designs, the detector system usually consists of a scintillation camera which can be found in nuclear medicine clinics. To collect the gamma photons emitted from the patient, the camera is fitted with collimator which consists of holes separated by lead septa. Conventional collimator design usually consists of holes with parallel septal walls. The parallel hole design provides a one-to-one ratio between the spatial extent of the radioactivity distribution and the detected image. The design parameters, such as the size, shape and length of the collimator holes, and the septal thickness, determine the acceptance angle of photons that hit the scintillation crystal and consequently the counting statistics and spatial resolution of the detector system.

Since collimator design has direct effects on the spatial resolution and statistical noise fluctuations in the detected image, it is a subject of importance in conventional nuclear medicine and SPECT imaging. Ideally, we would like to a design collimator with the highest detector efficiency with the best spatial resolution. Unfortunately, for a particular geometric design, such as the parallel-hole collimator, the increase in detection efficiency is inversely proportional to improvement in spatial resolution. That is, increase in detection efficiency must be accompanied by degradation in spatial resolution. This constraint imposes limits on the quality of conventional nuclear medicine and SPECT images.

New collimator designs have been used in SPECT imaging to improve the trade-off between detection efficiency and spatial resolution compared with the traditional parallel-hole collimators. Specifically, the new designs consist of collimators with converging holes. Examples are fan beam (*3, 4*), cone beam (*5, 6, 7*) and astigmatic collimators. These collimators are designed to image small parts of the body such as the head by magnifying the object of interest to fill the entire scintillation crystal area. The result is a substantial increase in detection efficiency for the same spatial resolution as compared to collimators with parallel holes. Typically, a fan beam and a cone beam collimator can be designed to have detection efficiencies on the order of 1.5 and 2.3 of that of a parallel-hole collimator with the same spatial resolution, respectively. Alternately, the increase in detection efficiency can be traded for improved spatial resolution. Since the trade-off is limited field-of-view, these collimators have proven useful in SPECT imaging of the head.

Recently, we have studied the use of fan beam and cone beam collimators for cardiac SPECT.(*8*) By fitting the heart within the field-of-view, we can take advantage of the improved trade-off between detection efficiency and spatial resolution of the converging-hole collimator design. A problem in the application of the converging-hole collimator designs in cardiac SPECT is truncation of projection data which lie outside the field-of-view of the collimator. For the typical detector rotation around a fixed central axis, another problem is the limited number of projection views for reconstructed image planes which are distant from the midplane of the collimator. When the limited angular projection data are used in a filtered backprojection algorithm, the reconstructed images are blurred along the direction of the central axis.

We are investigating the use of converging-hole collimators in cardiac SPECT imaging for improved clinical diagnosis. For fan beam collimators the truncated projection problem can be compensated for by extrapolating the projection data or by iterative reconstructed methods. Study of the limited angular reconstruction problem is underway to study its effects in lesion detection. These studies will provide understanding of the clinical efficacy of the new collimator designs in cardiac SPECT.

COMPENSATION AND IMAGE RECONSTRUCTION METHODS

Aside from statistical noise fluctuations, the projection data in SPECT is degraded by the finite spatial resolution of the detector, and photon attenuation and scatter in the patient. Compensation for the attenuation, detector response function, and scatter results in improved image quality and quantitative accuracy in SPECT and has been the subject of intensive investigation.

Most of the compensation methods are based on some assumptions such that the problem can be solved in closed analytic form. For example, if we can assume that the attenuation throughout the cross sectional image is uniform, e.g. the head or abdomen region, compensation for attenuation can be achieved using simple techniques which are derived analytically.(*9, 10, 11, 12*) Furthermore, if we can assume spatially invariant detector and scatter response functions, compensation for the detector response function can be achieved through the use of deconvolution techniques such as restoration filtering, i.e. the Wiener or Metz filters.(*13, 14*)

However, the assumptions of uniform attenuation distribution and shift invariant detector and scatter response functions do not apply universally. For example, in cardiac SPECT imaging, the attenuation coefficient of the lung region is one third of that of the surrounding muscle. Here, the assumption of uniform attenuation is invalid. The spatial resolution of a collimator-detector in a SPECT system degrades as function of distance from the detector. Contributions from scatter add to the complexity of the total detector response function whose shape depends on the position of the radioactive source, and geometry and composition of the scattering medium.(*15, 16, 17*) The variation of the total detector response violates the assumption of shift-invariant response function used in image processing.

We are investigating compensation methods for non-uniform attenuation distribution, and shift-invariant detector response function in cardiac SPECT imaging. Since conventional compensation and reconstruction methods do not apply in these situations, we are studying the use of iterative reconstruction methods in the compensation scheme.

In an iterative reconstruction method, an initial estimate of the object distribution is assumed. Reprojection data are generated from the initial estimate and compared with the corresponding measured projection data.(*18*) The differences between the reprojection and measured projection data are backprojected into an error image which is then used to update the initial estimate to obtain a new estimate or reconstructed image. Different update schemes are used in different iterative algorithms. The reprojection and backprojection operations are repeated until the reconstructed image approaches the true object distribution.

The physical factors which affect SPECT imaging can be modeled in the projector and backprojector used in the iterative reconstruction algorithm. We have developed projectors and backprojectors which model the exact attenuation distribution of the patient and/or the response function of the detector used in the SPECT imaging system. These projectors and backprojectors are used in the iterative reconstruction algorithms to compensate for image degrading factors.(*19, 20, 21*)

We have evaluated the effectiveness of the corrective reconstruction methods in cardiac SPECT imaging using simulation and clinical data. Figure 1 shows the cardiac phantom used in the simulation study. The phantom was generated based on an X-ray CT image to realistically simulate patient data. Projection data which include the effects of attenuation, detector response, scatter and noise were generated from the phantom. Reconstructed images were obtained by applying different reconstruction methods to the projection data.

Figure 2 shows a comparison of the reconstructed images obtained from different reconstruction methods. No statistical noise was added to the projection data. The reconstruction methods shown are the conventional filtered backprojection, iterative ML-EM algorithm with compensation for attenuation, and iterative ML-EM algorithm with compensation for both attenuation and detector response. The comparison demonstrates that the reconstructed image obtained using the conventional filtered backprojection method is of poor quality in terms of spatial resolution, quantitative accuracy, and image artifacts. The reconstructed images obtained with the iterative ML-EM algorithm are in general superior. Specifically, the reconstructed image with attenuation compensation provides greater quantitative accuracy. When compensation for both attenuation and detector response is included, the reconstructed image exhibits better spatial resolution together with greater quantitative accuracy.

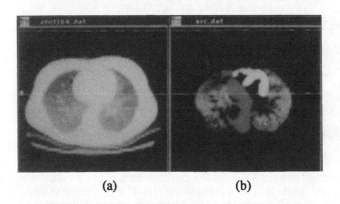

(a)   (b)

Figure 1. Simulated cardiac-chest phantom. (a) X-ray CT image obtained from a patient study showing the attenuation distribution. (b) Simulated ^{201}Tl uptake in the myocardium and lungs. The ratio of the activity concentration in the two region is 6:1. The images are displayed in logarithmic scale to enhance the features at the low activity level.

Figure 3 shows a comparison of the same reconstruction methods with scatter and noise added to the projection data. Again, the comparison demonstrates the better image quality and quantitative accuracy obtained with the iterative ML-EM algorithm with compensation for attenuation and detector response compared with that obtained with the conventional filtered backprojection methods without any compensation.

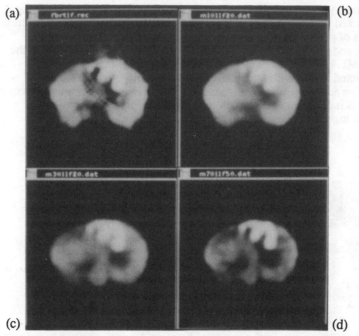

Figure 2. Reconstructed images of the cardiac-chest phantom shown in Fig. 1. The simulated projection data include the effects of attenuation and detector response, and is noise-free. (a) Filtered backprojection method without any compensation. (b) Iterative ML-EM algorithm after 20 iterations without any compensation. (c) Iterative ML-EM algorithm after 20 iterations with compensation for attenuation. (d) Iterative ML-EM algorithm after 50 iterations with compensation for both attenuation and scatter.

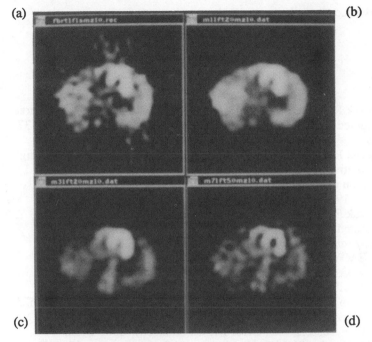

(a)

(b)

(c)

(d)

Figure 3. Reconstructed images of the cardiac-chest phantom shown in Fig. 1. The simulated projection data include the effects of attenuation, detector response, scatter, and Poisson noise fluctuations. (a) Filtered backprojection method without any compensation. (b) Iterative ML-EM algorithm after 20 iterations without any compensation. (c) Iterative ML-EM algorithm after 20 iterations with compensation for attenuation. (d) Iterative ML-EM algorithm after 50 iterations with compensation for both attenuation and scatter. All images have been filtered with a Metz filter with order equal to 10.

DISCUSSION

SPECT imaging is an important diagnostic tool for cardiac disease. The quality of cardiac SPECT images is degraded by physical factors which are difficult to compensate for using conventional methods. We have developed compensation methods by modeling the physical effects in the projector and backprojector of iterative reconstruction algorithms. These compensation methods showed improvement in terms of image quality and quantitative accuracy when compared with conventional reconstruction methods.

REFERENCES

1. Jaszczak RJ, Coleman RE Lim CB. SPECT: Single Photon Emission Computed Tomography. *IEEE Trans. Nucl. Med.* 1980; NS-27:1137-1153.
2. Jaszczak RJ, Coleman RE Whitehead FR. Physical Factors Affecting Quantitative Measurements Using Camera-Based Single Photon Emission Computed Tomography (SPECT). *IEEE Trans. Nucl. Science* 1981; NS-28:69-80.
3. Jaszczak RJ, Chang L-T Murphy PH. Single Photon Emission Computed Tomography Using Multi-Slice Fan Beam Collimators. *IEEE Trans. Nucl. Sci.* 1979; NS-26(1):610-618.
4. Tsui BMW, Gullberg GT, Edgerton ER, et al. Design and Clinical Utility of a Fan Beam Collimator for SPECT. *J. Nucl. Med.* 1986; 27:810-819.
5. Feldkamp LA, Davis LC Kress JW. Practical cone-beam algorithm. *J. Opt. Soc. Am. A* 1984; 1:612-619.
6. Floyd CE, Jaszczak RJ, Greer KL Coleman RE. Cone Beam Collimation for SPECT: Simulation and Reconstruction. *IEEE Trans. Nucl. Sci.* 1986; NS-33:511-514.
7. Jaszczak RJ, Greer KL Coleman RE. SPECT Using a Specially Designed Cone Beam Collimator. *J. Nucl. Med.* 1988; 29:1398-1405.
8. Gullberg GT, Zeng GL, Christian PE, et al. Emission Computed Tomography of the Heart Using Cone Beam Geometry and Noncircular Detector Rotation. Proceedings of the XIth Information Processing in Medical Imaging (IPMI) International Conference. Berkeley, CA, 1989. in press.
9. Chang L-T. A Method For Attenuation Correction in Radionuclide Computed Tomography. *IEEE Trans. Nucl. Science* 1978; NS-25:638-643.
10. Sorenson JA. Quantitative Measurement Of Radioactivity In Vivo By Whole-Body Counting.
11. Tsui BMW, Chen C-T, Yasillo NJ, et al. A Whole-Body Scanning System For Collection of Quantitative In Vivo Distribution Data in Humans. 3rd International Radiopharmaceutical Dosimetry Symposium. 1981.
12. Gullberg GT Budinger TF. The Use of Filtering Methods to Compensate for Constant Attenuation in Single-Photon Emission

Computed Tomography. *IEEE Trans Bio Eng* 1981; BME-28:142-157.

13. King MA, Schwinger RB, Doherty PW Penney BC. Two-Dimensional Filtering of SPECT Images using the Metz and Wiener Filters. *J.Nucl.Med.* 1984; 25:1234-1240.
14. King MA, Penney BC Glick SJ. An image-dependent Metz filter for nuclear medicine. *J. Nucl. Med.* 1988; 29:1980-1989.
15. Floyd CE, Jaszczak RJ, Harris CC, et al. Monte Carlo evaluation of Compton scatter subtraction in single photon emission computed tomography. *Med.Phys.* 1985; 12:776-778.
16. Frey EC Tsui BMW. Parameterization of the Scatter Response Function in SPECT Imaging Using Monte Carlo Simulation. *IEEE Trans. Nucl. Sci.* 1990; in press:
17. Rosenthal MS Henry LJ. Scattering in Nonuniform Media. *Phys. Med. Biol.* 1990; 35:265-274.
18. Lange K Carson R. EM Reconstruction Algorithms for Emission and Transmission Tomography. *J. Comput. Assist. Tomogr.* 1984; 8:306-316.
19. Gullberg GT, Huesman RH, Malko JA, et al. An Attenuated Projector-Backprojector for Iterative SPECT Reconstruction. *Phys.Med.Biol.* 1985; 30:799-816.
20. Tsui BMW. Implementation of Simulataneious Attenuation and Detector Response Correction in SPECT. *IEEE Trans. Nucl. Sci.* 1988; 35:778-783.
21. Tsui BMW, Gullberg GT, Edgerton ER, et al. Correction of Nonuniform Attenuation in Cardiac SPECT Imaging. *J. Nucl. Med.* 1989; 30:497-507.

V.

ENGINEERING ASPECTS
OF COGNITIVE SCIENCES

"Simple" Analog Circuits for the VLSI of Neural Networks

Simon Y. Foo and Lisa R. Anderson

Department of Electrical Engineering
FAMU/FSU College of Engineering
Florida State University
Tallahassee, FL 32306
(904)487-6474

ABSTRACT

The very-large-scale-integration (VLSI) of artificial neural networks (ANNs) requires "simple" processing elements and resistive interconnection circuits with high packing capabilities. However, most of the electronic components developed for ANNs are complex high-precision devices that lack VLSI capabilities, and therefore are not suited to implementing large-scale neural network systems of practical interests. Since ANN models are inherently robust and adaptive, it is possible to implement electronic neural networks using simple analog components with moderate precision and operate within tolerable noise margins. We will explore some of the traditional and novel techniques for building linear/nonlinear devices to simulate the synaptic weights and processing elements of ANNs.

1. INTRODUCTION

Artificial neural networks (ANNs) are systems which attempt to mimic, at least partially, the structure and functions of brains and nervous systems. There are billions of biological neurons in our brains interconnected in a manner whereby we can reason, memorize, compute, and so forth. Recently, there has been a strong resurgence in the research of neural computing for real-time applications due to the advancement in VLSI technology and optics. Current artificial intelligence (AI) technology based on knowledge-based expert systems has relied heavily on symbolic manipulations. The major limitation of the AI approach is that the knowledge base is a static set of rules cast by human experts. Thus, there exists an error-prone interface between the AI programmers and the human experts, where the programmers have to cope with fuzzy information.

Neural networks, on the other hand, are trained successively "by examples" in a real-world environment where they develop their own rules internally while adapting to changes in the environment. One of the advantages of ANNs is the ability to handle fuzzy or incomplete data. In particular, ANNs employ an enormous number of communication links among the processing elements (PEs) to perform distributed parallel processing (PDP). Due to the robust (or fault-tolerant) nature of ANNs, a few degraded or nonfunctional PEs will not greatly affect the overall neural network operation. Therefore, the speed and robustness of ANNs make them very attractive for a variety of applications such as pattern recognition, robotic control, and combinatorial optimization.

2. BUILDING BLOCKS

An artificial neuron can be modeled as a multi-input nonlinear thresholding device with weighted interconnections (synapses). The cell body of an electronic neuron is represented by a nonlinear amplifier (e.g., a high-gain amplifier), while the synapses are represented by variable resistors, as shown in Figure 1. The dynamics of each neuron is governed by an ordinary first-order differential equation (or difference equation for discrete-time systems) which describes the motion of the neural network.

Signals received from other neurons in the form of potentials across the resistive interconnects are collected by summing currents. Each synaptic weight or resistive interconnect is modeled as a passive resistor with conductance G. Based on its input neural voltages U, the PE produces an output signal V according to its nonlinear transfer function f, and the output signal V is then propagated to other neurons.

2.1. Variable Linear Conductance Devices

One of the most important aspects of neural networks is their learning capability, whereby synaptic strengths between neurons are adaptively changed according to an algorithm. Such learning could be supervised (e.g., Hopfield's) or unsupervised (e.g., Kohonen's). From neurobiology, we know that a human brain has about 100 billion neurons, and that each neuron (or nerve cell) is typically connected to approximately 10,000 other neurons. Since biological neurons respond only at the millisecond time scale (much slower than transistors), it is apparent that the computation abilities of our brains arise from the large number of neurons and the huge number of interconnection links. Due to the physical limitation of a two-dimensional silicon wafer, electronic neural networks have to rely heavily on "simple" models of neurons and synapses.

As discussed earlier, synaptic weights can be simulated by variable linear resistors. Perhaps the simplest active devices for simulating a variable resistor are the junction field-effect transistors (JFETs) and metal-oxide-semiconductor field-effect transistors (MOSFETs). Both the JFET and the MOSFET act as voltage-controlled linear resistors for small values of V_{ds} (drain-to-source voltage), operating in the active region. The output resistance of the device is controlled by the input gate-to-source voltage, V_{gs}, and as V_{gs} increases, the resistance decreases. The MOSFET has been shown to behave as a linear resistor if the output voltage lies between approximately -0.5 volt to +0.5 volt. This is a feasible range since we intend to keep the operation of the neural network in the millivolts so as to reduce power dissipation. The depletion-mode device has the same characteristics as the enhancement-mode device except that V_{gs} is negative.

Similarly, an n-channel JFET also acts as a voltage-controlled linear resistor for small values of V_{DS}. The linear region lies approximately from -1.0 volt to +1.0 volt. Thus the JFET has a wider dynamic range than the MOSFET.

It is also possible to utilize MOS switches and floating active resistors for simulating variable linear resistors. Both devices can be implemented using a single MOSFET. However, a more practical approach is the CMOS switch based on a pair of complementary MOSFETs, as shown in Figure 2. The CMOS switch is essentially a transmission gate used in digital circuits. The resistance of the CMOS switch is controlled by the voltage at the bulk of transistor $M1$ and the gate voltage of $M2$. The linear resistive region lies approximately from -1.0 volt to +1.0 volt. In our example shown in Figure 2, the output resistance is approximately equal to 3.7 kΩ. The CMOS switch has a major advantage over a single MOSFET switch or a floating active resistor. In particular, the problem of clock feedthrough is eliminated through the parallel connection of the n- and p-channel devices, which require opposing clock signals. Consequently, the dynamic range is greatly increased as a result of the complementary devices.

In our previous approach, we proposed the analog variable resistor techniques for simulating the synaptic strength [1]. Basically, there are four types of a variable resistor: a switched-capacitor circuit, a switched-resistor, a switched-ladder resistor, and a voltage-controlled resistor. The purpose of these devices is to linearly vary the flow of current controlled by a clock pulse or an input voltage.

In a switched-capacitor circuit, the value of resistance R depends on an input clock frequency f_{clk} and a capacitor C, where

$$R = 1/(f_{clk} \ C) \ .$$

As the clock frequency and capacitance increase, the resistance decreases.

A switched-resistor circuit is composed of a fixed R_o resistor, an analog switch, and a capacitor. The value of the resistance R is determined by the ratio

$$R = R_0 \ / \ d \ .$$

where d is the duty cycle of the switch. As the duty cycle decreases, the value of R increases.

A switched-ladder resistor is composed of n analog switches in parallel with n R_o to $2^{n-1}R_o$ resistors in series. The total resistance is controlled by the analog switches. As a result, there are $(2^{n+1} - 1)$ possible values ranging from 0 ohms to $(2^{n+1} - 1)R_o$ ohms. For example, if all of the switches are turned "off", the total resistance from one end to the other becomes $(2^{n+1} - 1)R_o$ ohms.

Analog devices with high accuracy can be built at the expense of larger silicon area and higher power dissipation. For example, we could utilize an elaborate and precision voltage-controlled linear resistor introduced by Czarnul [2]. This circuit based on a matched pair of FETs. The linear resistance is determined by the gate voltage V_{G2} of transistor T_2 and the floating voltage source V_c of transistor T_1. The output current I_o is determined by

$$I_o = \beta \ V_{in} \ (V_c - V_{G2})$$

where β is the transconductance parameter dependent on the fabrication and the geometry of the transistors.

Similar high-precision but "expensive" approaches include a CMOS voltage-controlled linear resistor with a wide dynamic range by Youssef, Newcomb, and Zaghloul [3]. A pair of complementary enhancement-mode MOS transistors is used to offset the nonlinearity present in the circuit. Although this circuit provides high resolution resistance with a wide dynamic range, the circuit requires several external voltage sources which makes it expensive to implement. This overhead is hardly justified since neural networks are fault-tolerant requiring only moderate accuracy.

Other novel approaches to analog interconnects include an electrically trainable network with 10,240 synapses which uses "floating gate" non-volatile memory for analog storage of synaptic weights [4], developed by researchers at Intel. It is also possible to utilize charge-coupled devices (CCDs) and metal-nitride-oxide-silicon (MNOS) devices for storing synaptic weights. The CCDs are arrays of metal-insulator-silicon (MIS) capacitors with packets of charge which can represent synaptic excitation. These packets of charge are formed, transferred from place to place, and collected by varying the voltages on the gate, thus providing a means of communication and computation. MNOS devices are MIS capacitors with a dual dielectric comprising a very thin layer of silicon dioxide "sandwiched" between the main silicon nitride insulator and the silicon substrate. MNOS devices provide analog non-volatile storage by allowing the carriers to pass through the oxide layer at high gate voltages (through quantum-mechanical tunneling between the silicon and the traps in the nitride) and stopping carrier passage at lower gate voltages, trapping the carriers in the nitride. During the recall (or evaluation) phase of neural network operation, MNOS devices can be used to control the operations of CCDs. To complement the evaluation phase, MIT's Lincoln Laboratory has developed a mechanism allowing CCDs to control MNOS devices during the learning phase of neural network operation.

2.2. Deterministic Nonlinear Processing Elements

The processing elements (PEs) are the other key components of a neural network. The PEs collect and process all the incoming signals propagated from other neurons through the synapses. Based on a nonlinear activation function, the neuron may "fire" if the sum of input signals exceeds a certain threshold, or it may be turned "off", otherwise.

In general, there are three basic nonlinear transfer functions for artificial neurons, i.e., high-gain limit, linear threshold, and sigmoid. The high-gain limit (or step) function of binary or two-state neurons for configuring associative memory in the Hopfield networks, Hamming networks, and Boltzmann machines can be easily implemented by an analog comparator. The threshold of the analog comparator is controlled by the reference voltage V_{ref}.

The linear threshold activation function is commonly used in the single-layer feedforward networks such as ADALINE and single-layer perceptrons. The linear function with a fixed threshold can be easily implemented by a noninverting operational amplifier circuit. A twin cascaded inverting amplifiers can be used to implement a linear function with adjustable threshold.

Another nonlinear function widely used in the sigma-pi and Hopfield neural networks is the continuous sigmoid function. In multi-layer feedforward associative networks (or backpropagation networks), the learning process depends on the delta rule [5] which requires a differentiable and nondecreasing function such as the sigmoid transfer function:

$$V = 0.5 \ \tanh(\lambda U)$$

where λ is the voltage gain, and U and V are input and output voltages, respectively. A sigmoid function with a fixed gain can be realized by a high-gain inverting amplifier in cascade with a unity gain inverting amplifier. It is also possible to utilize digital devices to construct a sigmoid circuit. For example, the sigmoid neuron (with normal and inverted outputs) used in Hopfield network can be "approximately" implemented by twin cascaded CMOS inverters, as shown in Figure 3. The voltage gain (gain of sigmoid amplifier) is then controlled by the aspect (W/L) ratios of the transistors.

Yet another useful sigmoid circuit with variable gain control uses an unbuffered voltage comparator with a positive feedback loop and double cascaded inverters (output stages) with negative feedback loops. The unbuffered comparator provides an approximate sigmoid transfer function, while the negative feedback loops of the inverters act as gain controls. Figure 4(a) shows a detailed schematic of the variable-gain sigmoid circuit. The gain of the sigmoid function can be increased by increasing the ratio $r = R_8/R_7 = R_{10}/R_9$, i.e., the gains of the inverters. SPICE simulations of the variable-gain sigmoid circuit based CMOS operational amplifiers were performed for $r = 100$, 250, 500, and 1000. The results are plotted in Figure 4(b). It is observed that as r increases, the sigmoid curve approaches a high gain limit function; lowering r results in a sigmoid curve with gentler slope.

3. CONCLUSION

In general, a biological neuron can be modeled as a very high fan-in/fan-out processing element, whereby a neuron is typically connected to several thousands of neurons. Consequently, the wiring of the large number of resistive interconnects on a two-dimensional surface of a silicon wafer represents the major bottleneck for implementing electronic neural networks.

In this paper, we have briefly described several analog circuits for the VLSI implementations of neural networks based on discrete and integrated devices. We believe that the moderate resolution of these "simple" analog components are sufficient for implementing Hopfield's model and Fukushima's Neocognitron paradigm. More details on the analog circuits can be found in a recent paper by Foo, Anderson, and Takefuji [6].

4. REFERENCES

[1] S. Y. Foo, Y. Takefuji, and T. J. Harrison, "Analog Components for Electronic Neural Networks," *1st Florida Annual Microelectronics Conference*, Boca Raton, FL, May 10-11, 1989.
[2] Z. Czarnul, "Design of Voltage-Controlled Linear Transconductance Elements with a Matched Pair of FETs," *IEEE Trans. on Circuits and Systems*, CAS-33, No. 10, 1986.
[3] H. Youssef, R. Newcomb, M. Zaghloul, "A CMOS Voltage-Controlled Linear Resistor with a Wide Dynamic Range," *Proc. of 21st Southeastern Symposium on System Theory*, Tallahassee, FL, March 26-28, 1989.
[4] M. Holler, S. Tam, H. Castro, and R. Benson, "An Electrically Trainable Artificial Neural Network (ETANN) with 10240 Floating Gate Synapses," *Proc. of IJCNN*, Washington, DC, June 1989, Vol. II, pp. 191.
[5] J. L. McClelland, D. E. Rummelhart, and the PDP Research Group, "Parallel Distributed Processing, Volumes I and II," *MIT Press*, 1986.
[6] S. Y. Foo, L. R. Anderson, and Y. Takefuji, "Analog VLSI Components for Neural Networks," *IEEE Circuits and Devices Magazine*, July 1990.

Suggested Further Reading:

[1] H. P. Graf and L. D. Jackel, "Analog Electronic Neural Network Circuits," *IEEE Circuits and Devices Magazine*, July 1989.
[2] C. Mead, "Analog VLSI and Neural Systems," Addison-Wesley, 1989.
[3] N. H. Farhat, "Optoelectronic Neural Networks and Learning Machines," *IEEE Circuits and Devices Magazine*, September 1989.
[4] L. E. Atlas and Y. Suzuki, "Digital Systems for Artificial Neural Networks," *IEEE Circuits and Devices Magazine*, November 1989.

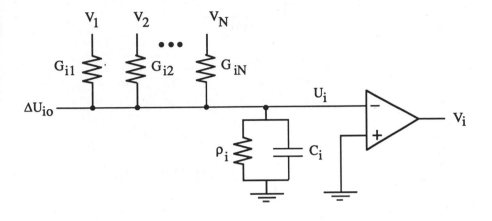

Figure 1. Circuit schematic of an artificial neuron.

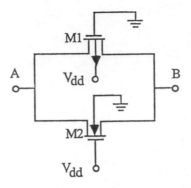

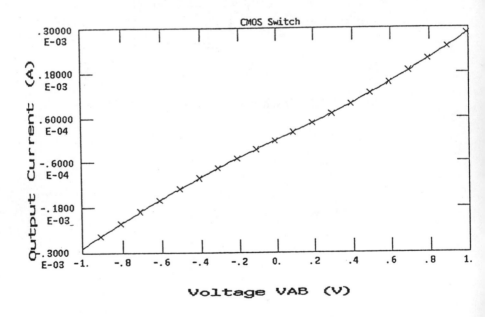

Figure 2. A CMOS switch and its characteristics in the active region.

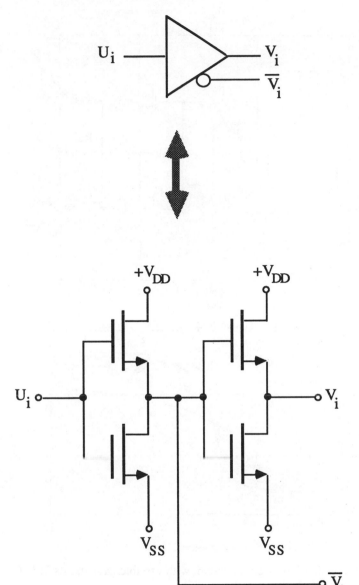

Figure 3. Twin cascaded CMOS inverters to perform sigmoid transfer function. The output voltage gain can be adjusted by adjusting the aspect ratios of the transistors.

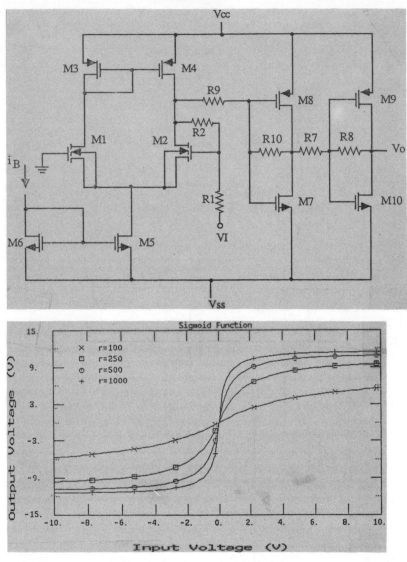

Figure 4. Schematic of the sigmoid circuit with variable gain and its SPICE outputs for various *r*'s.

ADAPTIVE LEARNING AND NON-IDEALITIES IN NEURAL NETWORKS

Robert C. Frye, Edward A. Rietman and Chee C. Wong

AT&T Bell Laboratories

Murray Hill, NJ 07974

ABSTRACT

Experimental results of adaptive learning in an optically controlled neural network are presented. These networks mimic an architecture commonly found in biological systems. One of their more interesting properties, shared by their biological counterparts, is their ability to adaptively learn despite nonidealities such as component variations and failures. These experiments offer insights into the role of adaptation in obtaining high levels of performance from non-ideal components.

INTRODUCTION

The underlying architecture of neural networks has resulted from the study of models of biological neural systems. Originally, these models were developed in an attempt to understand and explain some of the uncommon processing capabilities of biological networks. It has since become apparent that artificial networks based on these models exhibit many of their biological counterparts' useful and remarkable properties, including the ability to adaptively learn, to self organize and to generalize the solutions that they derive from their training examples to new problems outside their past experience.

Figure 1 illustrates an artificial neural network of the kind that we have studied. These networks generally consist of layers of active elements, the *neurons*, coupled together by variable strength connection matrices simulating the *synapses* found in biological networks. In the figure, the rectangles indicate neurons and the lines connecting each node

241

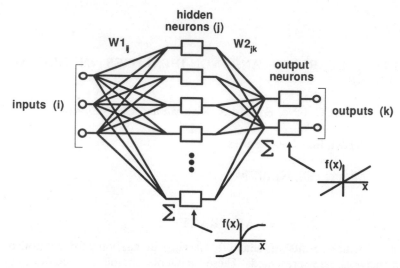

Figure 1. Layered feed forward neural network. The rectangles represent neurons and the connecting lines represent variable strength synaptic weights.

indicate synaptic connections. In analog electronic hardware versions of these networks, this highly parallel architecture allows them to operate at a high equivalent computational rate. Software networks, based on this physical model, do not realize the same benefits of parallelism, but can still exploit the ability of these networks to adaptively learn. The algorithms that have been developed to design neural networks that do particular computational tasks can be applied to both hardware and software networks.

The interconnections between the layers of neurons are variable connection strength coefficients in a weight matrix. In a software network, the physical architecture shown in Figure 1 can be described by the equation

$$o_k = \sum_j W2_{jk} \tanh \left[\sum_i W1_{ij} \, i_i \right]$$

This relation describes o_k, the output of the k*th* neuron in the output layer, for a set of inputs, i_i. The matrix W1 is the input-to-hidden layer matrix and the W2 matrix is the hidden-to-output matrix.

Notice that apart from the number of layers and the number of neurons in each layer, the basic architecture and elements within the network are generic. One network of this kind looks more or less like any other. The function of the network is not determined by its active elements, the neurons, but rather by the interconnections among them. The basic idea behind the adaptive process is to adjust the values of the individual connections $W1_{ij}$ and $W2_{jk}$ to minimize the rms error in the output. This process is done gradually - *i. e.* the changes to the connections after each individual trial are small. With each change, however, the network more closely approximates the desired response.

A widely used adaptive method to compute the changes in the connection matrices is the back-propagation of errors technique, as discussed by Rumelhart, *et al.* [1]. This method is a generalization of the delta rule developed by Widrow and Hoff [2] and is based on a gradient descent optimization technique. It attempts to minimize the mean-squared error in the output of the network, as compared to a desired response. This technique results in lowering the average error as the weight matrices in the network evolve. In layered networks of the kind that we will be discussing, the changes in the weights after each trial are proportional to the error itself. This leads to a system that settles to a stable weight configuration as the errors are minimized. The final configuration may result in a true global minimum or only a local one, but from a practical standpoint the gradient descent algorithm generally results in useful, if not optimal, solutions. (Several variations to this back-propagation approach have been used to address some of its deficiencies. These are reviewed by Jacobs [3].)

The adaptive technique described above has proven to be successful in digital software based networks. Implementing the same techniques in analog electronic hardware, however, is more difficult. Truly adaptive hardware requires the fabrication and interactive control of large numbers of variable resistance interconnections. By far the most successful method currently available for fabricating and interconnecting large numbers of electronic components is VLSI. An attendant consequence of this technology, however, is the surprisingly large component-to-component variation that is typical of VLSI devices [4]. These variations, which have a negligible effect on digital circuits, can be particularly troublesome in the fabrication of analog circuits based on resistive arrays, since the circuit function depends critically on the component values. Adaptive techniques are viewed as a possible way to build analog VLSI networks that are not overly constrained by these problems of component variations.

ADAPTIVE HARDWARE

Using optically programmable synapses and the electronic neurons, we designed and built a network with the layered feed-forward architecture of Figure 1. A block diagram of the system is shown in Figure 2. The network had three analog inputs, and its neurons (ten hidden nonlinear neurons and two linear output neurons) were built using conventional operational amplifiers. The variable resistance synapses were built using an array of amorphous silicon photoconductors whose resistances were modulated with a projected image. A digital computer generated the training data, evaluated the output and the error, and updated the image that programmed the conductance values in the interconnection array. Inputs to the network were analog voltage levels, and the output of the network, and of the hidden neurons, were monitored and used in the weight-update calculations. (Details of this network and its construction can be found in reference [5].)

Measured values of the synaptic conductance in this hardware showed variations in the individual synapses, including the effects of device nonuniformities, optical misalignments etc, to be ± 30%. Moreover, these networks are prone to temporal noise. None of these non-ideal effects are present in software based networks, and the purpose of these experiments was to determine the extent to which they limit the network's ability to adaptively learn and perform.

FAULT TOLERANCE

We have successfully used this hardware to adaptively learn several simple nonlinear computations. This procedure consists first of choosing

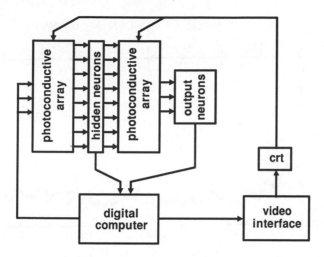

Figure 2. Block diagram of the optically controlled adaptive neural network.

an appropriate set of nonlinear equations linking two of the inputs to the two outputs. (The third input is set to a constant to allow the network, if necessary, to generate nonzero outputs for zero input). A familiar example of such a pair of equations is the transformation from polar to rectangular coordinates. Randomly generated inputs, together with the correct desired output are repeatedly presented to the network, and the error in the network's output is used to compute changes in the strength of the synaptic weights. After each trial, the weights are changed and the next trial is presented. If the network successfully learns, it's average error will fall with increasing numbers of trials.

Figure 3 shows the results of one such learning sequence. In this experiment, the network adaptively learned to compute the position of a ballistic trajectory 10 seconds after the object was tossed into the air with random initial velocity and angle. (This is the electronic analog of catching a ball. The neural network does not solve equations, but rather learns from experience where the ball will go based on the initial conditions.)

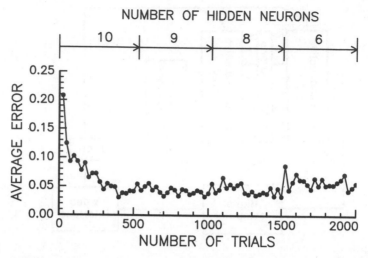

Figure 3. Learning curve for a simple ballistic predictor. At the indicated intervals, neurons in the hidden layer of the network were disabled.

One important conclusion that can be drawn from this experiment is that loss of neurons not directly associated with the output layer result in some loss of precision, but do not cause overall loss of functionality. This is partly a consequence of computational load sharing by the parallel neurons in the hidden layer. In addition, however, in this network, which is continually adapting, the network actually recovers after the sudden loss of one of its components. The computational load automatically gets redistributed among the remaining neurons.

COMPONENT VARIATIONS

A second, less obvious conclusion that can be drawn from the above experiment is that the network is performing at a level of accuracy that is higher than we would normally expect from a system built with such highly variable components (*i. e.* 5% relative error with 30% relative component variation). One explanation for this is that the variations are averaged out by the large number of synapses in the network. This explanation can be ruled out by a simple experiment. If we train a software network based on the average measured properties of the hardware, and then try to use the values of the synaptic weights learned in simulation to program the hardware network, we find that the networks performance is only slightly better than random guessing. (In the above example, this training method resulted in 30% relative error, compared with 33% for a random guess.)

Our experiments suggest that the ability of this highly non-ideal network to adaptively learn is a result of self-compensation for many of its internal non-idealities. This self-compensation appears to be an inherent property of the adaptive process, since it is not inherent in the learning algorithm, *i. e.* the algorithm is blind to component failure or nonuniformity. However, these non-ideal effects cause errors to occur. By adjusting the synaptic weights to minimize these errors, the network automatically compensates for its own weaknesses.

Figure 4 shows the results of simulations to determine the importance of these component variations in this network. These simulate variable amounts of component variation under two different conditions of adaptive learning. For data labeled "full feedback," the output of all neurons in the network were monitored and used in the weight update computations. For "partial feedback," on the other hand, only the two output neurons were monitored and the signal at the hidden neurons was inferred from the output and the presumed (*i. e.* average) values of the synaptic elements. As the component variations become larger, departure of the synaptic weights from their presumed values becomes greater, resulting in more uncertainty about the internal configuration of the net. Interestingly, with full feedback, component variations have no observable effect on the network's ability to learn, or on the final accuracy that it can obtain. Even with only partial feedback, the network can tolerate a surprisingly large degree of nonuniformity before it is adversely affected.

CONCLUSIONS

We have shown experimentally that the accuracy of an analog hardware neural network can significantly exceed that of its components. Moreover, these networks exhibit fault tolerance. Simulations aimed at exploring the importance of various nonideal effects show that they have an almost complete lack of sensitivity to large component variations. This is undoubtedly a consequence of the adaptive learning procedure

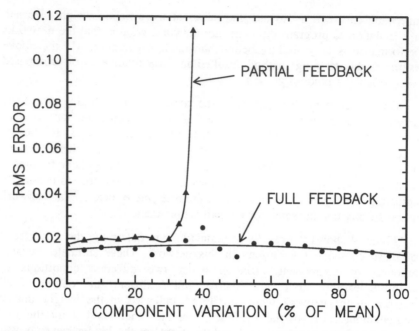

Figure 4. Simulation results showing the average error, after training, of networks as a function of overall variation in the synaptic elements. Partial feedback data were obtained by using only the values of the output neurons in the weight update computations, while for full feedback the hidden neurons were also monitored.

itself. The network, by learning in the presence of its internal faults and nonuniformities, automatically compensates for many of these problems. Although it is unlikely that biological neural networks use the same gradient descent minimization of errors method to adaptively learn, clearly on some level, trial and error play an important role in their adaptive process. These experiments demonstrate, at least in their limited context, that it is possible to obtain complex, accurate behavior from ensembles of nonuniform components, and may help to shed light on the capabilities and operation of biological networks.

REFERENCES

1. D. E. Rumelhart, G. E. Hinton and R. J. Williams, "Learning Internal Representations by Error Propagation" in *Parallel Distributed Processing: Explorations in the Microstructure of Cognition*, Vol 1, D. E. Rumelhart and J. L. McClelland (eds) MIT Press Cambridge MA (1986). also D. E. Rumelhart, G. E. Hinton and R. J. Williams, "Learning Representations by Back-Propagating Errors" Nature 323, 533 (1986).

2. B. Widrow and M. E. Hoff, "Adaptive Switching Circuits" IRE WESCON, Convention Record IRE, New York pp96-104 (1960).

3. R. A. Jacobs, "Increased Rates of Convergence Through Learning Rate Adaptation" Neural Networks, 1, 295 (1988)

4. M. A. Sivilotti, M. R. Emerling and C. A. Mead, "VLSI Architecture for Implementation of Neural Networks," AIP Conf. Proc. #151: *Neural Networks for Computing*, J. S. Denker (ed), American Inst. Physics, New York (1986).

5. R. C. Frye, E. A. Rietman, C. C. Wong and B. L. Chin, "An Investigation of Adaptive Learning Implemented in an Optically Controlled Neural Network," Proc. Int'l Joint Conference Neural Networks, II, 457, Washington DC (1989).

REFERENCES

NEURAL NETWORK APPLICATIONS IN

BIOMEDICAL ENGINEERING

D. Benachenhou, M. Cader, H. Szu*, and L. Medsker

Department of Computer Science and Information Systems
American University, Washington DC 20016

* Naval Research Laboratory
Washington DC 20375

ABSTRACT

Expert systems (ES) have been applied to the area of medical diagnosis with celebrated results such as those derived by EMYCIN. However, experience has shown that growth of such rule structures increase the complexity of verification and subsequent maintenance. We wish to ease this increasing complexity, while maintaining the attributes of an intelligent system, by using an integrated system. More specifically, it is evident that intelligent computer systems must be user-friendly, system adaptive, and trustworthy. A neural network can deal with open sets of input data derived by expert systems and use output which can be strengthened by the fault-tolerant recall of associative memory and the self-organizing property of the Adaptive Resonance Theory (ART). This integration of traditional expert system technology with neural network technology is evident in the system that was designed to distill a large set of candidate primers into a smaller set of effective PCR primers (or superset primers) that can amplify a segment of DNA infected by any of several possible AIDS viruses: HIV I-IV. This system also identifies important primer characteristics and superset primers which might be useful for medical doctors and laboratory technicians.

251

1. INTRODUCTION

Neural networks have been applied to biomedical engineering in several interesting areas. The first important area is that of refining medical knowledge bases by neural networks. For example, a semantic network consisting of O-A-V triplets of object, attribute, and value can be assigned a probability value and a certainty factor, thus becoming a belief net. Multiple layers of such networks are trained with a backward error propagation algorithm in order to find inter-dependency relations. Then, use of a labeling convention wherein the neurons are treated as vertices in a graph leads to the reading of heavily weighted nodes as data dependencies, subsequently interpreted as rules [LIMI01], [GALL01], and [JACO01].

The second important area is that of real-time signal processing for monitoring vital signs in an intensive care unit (ICU); e.g., those used at the McGill Research Center for Intelligent Machines. Such sophisticated medical equipment generates burdensome amounts of data that must be analysed in the temporally critical environment of an ICU. Thus, neural intelligent expert systems become applicable as knowledge filters that learn which features of the signal are important. In this area, we envision adaptive interfaces that are user-friendly to an extent dependent upon the level of the user's experience or work history. The third area is that of biomedical image processing for enhancement, recognition, and reconstruction of EEG, ECG and other sensor generated signals; and analysis of CT scan images of brain tissue, internal organs, and other parts of the human anatomy. Further applications, that use concepts from all three areas include dermatology diagnosis [YOON01], and medical diagnosis based upon symptom-disease-diagnosis tuples learnt from medical cases [LEVE01]. The fourth is the present area of non-geometric and non-parametric biomedical applications using a DNA related knowledge base in order to select primers to accomplish the polymerase chain reaction (PCR) [MULL01]. That is, the generation of pairs of primers, and primer features without a priori knowledge of the geometry and underlying statistics of the input space.

Generally the solution space to a particular problem may be only partially expressed within the rule structure of an ES. In such systems, a closed input problem space is necessary for arrival at valid decisions because the system is only cognizant of the problem representation as expressed within its rule base. However, many biomedical applications encompass a more dynamic input space in which the variability and the ranges of the input domain may not be completely specified at the time of system design. As such, an open environment with an immense data throughput rate and in which a priori relationships (statistical dependencies) amongst variables may be unknown might result in an ES arriving at a wrong or incomplete answer. The PCR-primer selection problem specification is that a valid set of candidate primers must satisfy a number of rule-based primer selection constraints, such that amplification (by PCR) of a HIV virus injected into any of a hundred thousand sites in the DNA may occur, is in the class of NP (non-deterministic polynomial time) problems.

2. THE INTEGRATED SYSTEM

The integrated system, consisting of both expert and neural network systems, produces primers that can accomplish the polymerase chain reaction of specific virally mutated DNA. In addition, novel properties of the primers emerge as a function of the feature extraction. With such feature or property extraction, we have been able to confirm the most probable rule (i.e., complementarity [BENA01]). Based upon the feature extraction, we are developing new rules to increase primer specificity for HIV infected DNA.

In figure 1, the expert system ❶ is used to apply empirically-derived rules to arrive at a possible set of candidate primers that we hope will amplify the desired DNA. However, with the high cost of primer synthesis, it is somewhat impractical to synthesize all the candidate primers and run trial-and-error experiments. An ART [CARP01] network measures the similarity of the primer input representations and clusters those with some leader primer template. Such a leader template is usually, but may not be limited to, a superset of all the input primer representations that cluster to it.

In addition, the ART network is designed to filter the input candidate primer space into a set of features that represent the invariant pattern elements (invariant bases) of all similar inputs. Such features may be interpreted as indicators for primer specificity to amplify DNA infected by certain HIV viruses.

There is no need for a priori knowledge of the underlying statistical distribution in the candidate primer space for proper clustering to take place. This is a partial result of the dynamics of the ART network, the concept of distributing the input primer over an array of neurons, and the input data representation encoding. In addition, it is not necessary for there to be a geometric representation of the input space such that similarity measures (eg., Euclidian) should be used. A variant of the dot product, that is most geometrically neutral, is used as a basis for the clustering of similar inputs.

3. SAMPLE CASES

In this section, we show the results from the rule base selection of primer candidates in Figure 2 (corresponding to stages ❶ and ❷ of Fig. 1). The self clustering property and leader properties that are derived by ART (corresponding to stages ❸ and ❹ of Fig. 1) are shown in Figures 3 and 4, respectively.

The sequence of bases that make up the primer is encoded using various schemes (one of which is shown in Fig. 2). Each input vector is then made incident upon the layer of input neurons forming a pattern of activity. The ART network then clusters and/or filters these inputs into primer features based on a dot product similarity measure. Although a stable memory state is derived, the dynamics of the ART system allows plasticity of the features and even addition to the memory state of the network. Such a property is highly desirable for the recognition of new strains of viruses that may occur (Figure 5).

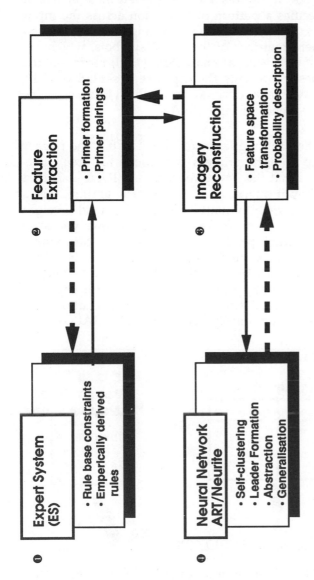

Figure 1. The solid arrows move clockwise from ❶ the Expert System where rules make up the constraints imposed upon the DNA sequence database. Application of the system derives pairs of candidate primers or features ❷. At ❸, the stage of Imagery Reconstruction, a transformation from the primer pair space into the feature space is assured by preservation of the most invariant elements (bases) and the self clustering property of ART ❹. Alternatively, the counter clockwise dashed arrows beginning at ❶ utilize a reconfigurable Hopfield network (Hairy neuron or neurite network) in order to generate both the solution space and ❷ use a probability controlled search procedure to find the global minimum (optimal solution). Abstraction of such minima (candidate primers) could lead to formation of rules for addition to the ES rule base ❷, ❶.

The following figures require some explanation so that they may be well understood. In figures 3 and 4, the left most matrix of forty "*" symbols specifies a pattern of activation across the input layer of neurons that corresponds to the encoding scheme outlined in figure 2. Subsequent matrices of forty "#" symbols, which are delimited by a line of five vertical characters ("|"), represent the output primer that best represents the input representation. The candidate primers have been derived from the virus sequence HIVJH34, a mutated sequence of viral DNA.

In Figure 3, when the vigilance (ρ) is set to 0.1, the primer candidate input space of eight primers is reduced to three. It is evident by these results that the first three primers (IN: 0, 1, 2) best represent the whole input space. In a laboratory setting such a savings, derived from having to synthesize three instead of eight candidate primers, would be most welcome. At higher vigilance levels; however, the network produces the degenerate case where no superset primers (or leaders) may be found. In other applications of the ART network, we have found that invariant properties of similar candidate primers provide us with clues for rule formation. However, such rule generation and base dependencies cannot be automatically generated as in [GALL01].

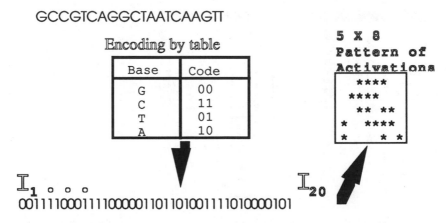

Figure 2. The input candidate primers are encoded as a 40 element binary vector. This is then presented to the ART network, forming a pattern of activations across the input neuron layer .

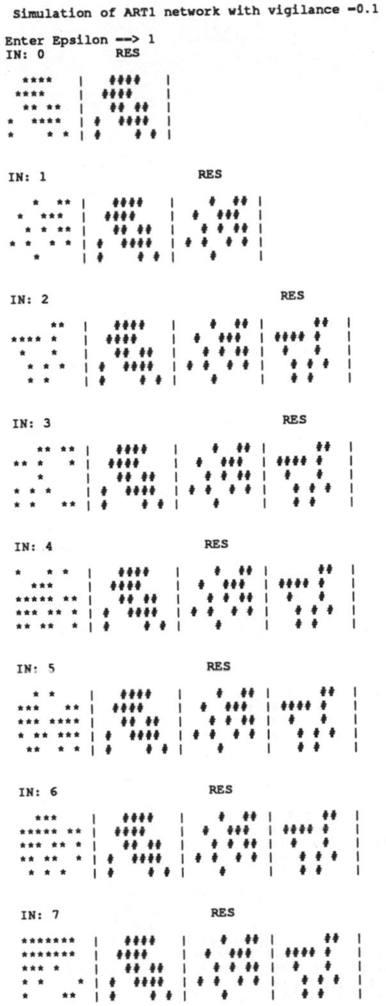
Simulation of ART1 network with vigilance =0.1
Enter Epsilon --> 1
IN: 0
RES
IN: 1
RES
IN: 2
RES
IN: 3
RES
IN: 4
RES
IN: 5
RES
IN: 6
RES
IN: 7
RES

Figure 3. The ES derived input space (left most column) is presented to the ART neural network in order to find the similarity amongst the input primer space. Those primers that are shown at stage (IN: 7) are the leaders that best represent the whole candidate primer space with a degree of similarity being allowed of 0.1.

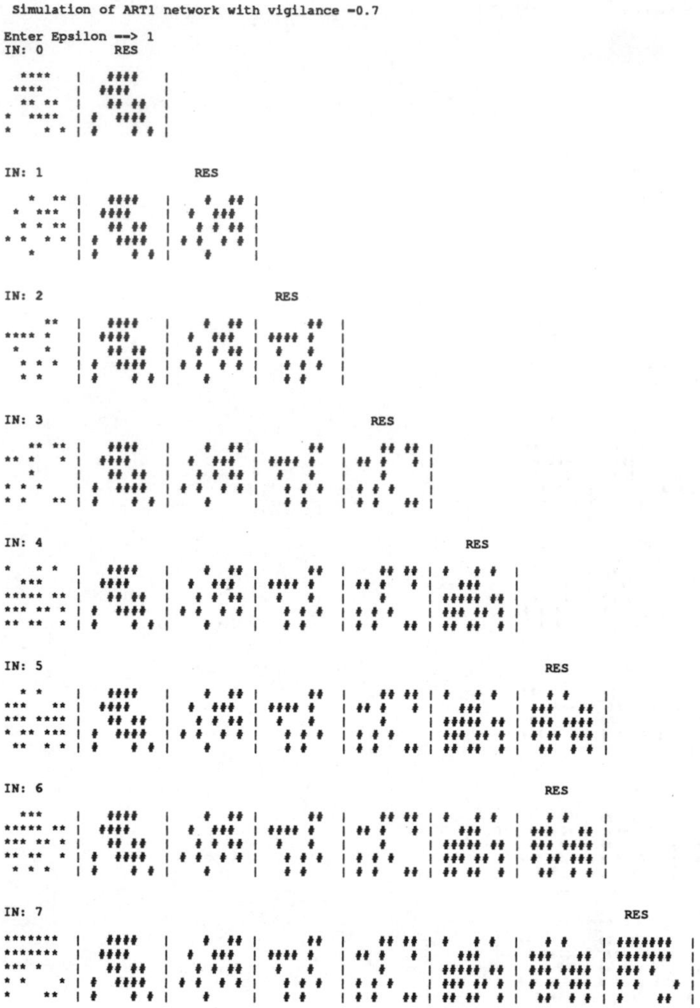
Simulation of ART1 network with vigilance =0.7
Enter Epsilon --> 1
IN: 0
RES
IN: 1
RES
IN: 2
RES
IN: 3
RES
IN: 4
RES
IN: 5
RES
IN: 6
RES
IN: 7
RES

Figure 4. The same candidate primer space is presented to the ART network as in Fig. 3; however, the degree of similarity that is necessary for clustering to occur is very small. As a result, the degenerate case , where each input candidate primer forms its own cluster occurs.

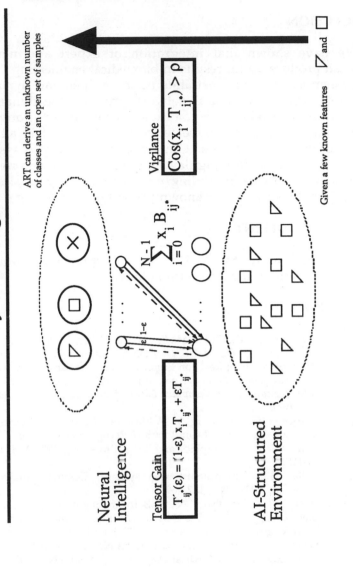

Figure 5. Illustration of the Hybrid ART network discovering new properties (X).

4. CONCLUSION

We have shown that integration of expert and neural systems can produce useful results in biomedical engineering. The expert system applies the constraints upon an open and dynamic environment, whilst the neural network provides the dynamics for valid decision making in a continually changing environment and the ability to generalize superset primers that contain all the properties of the subset primers. Otherwise, may the expert system may be described as performing flawlessly in a closed environment with the neural network providing fine enhancements of similar candidate primers such that unknown properties may emerge.

5. ACKNOWLEDGMENT

I would like to thank Dr. M. Z. Cader for valuable discussions concerning medical terminology.

6. REFERENCES

[BENA01]   Benachenhou, D., Cader, M., Szu, H. ,Medsker, L., Wittwer, C.,and Garling, D., "AIDS Viral DNA Amplification by Polymerase Chain Reaction Employing Primers Selected by AI expert system and an ART neural network", *Proceedings of Third IEEE Symposium on Computer Based Medical Systems*, Chapel Hill, June 3-6, 1990.

[CARP01]   Carpenter, G., Grossberg, S. "A Massively Parallel Architecture for a Self-Organizing Neural Pattern Recognition Machine." Computer Vision, Graphics, and Image Processing, 1987, Vol. 37, 54-115.

[GALL01]   Gallant, S.I. ,"Connectionist expert systems," Communications of the ACM, 31(2), 1988, 152-169.

[JACO01]   Jacob, R. J. K., Froscher, J. N., "A Software Engineering Methodology for Rule-Based Systems," IEEE Transactions on Knowledge and Data Engineering, 2(2), June 1990.

[LEVE01]   Levey, J., Brennan, C., Anderson, A., "A Neural Network Knowledge Base for Medical Diagnosis," J. Neural Network Computing, 2(1), 1990.

[LIMI01]   Li-Min Fu, "Integration of neural heuristics into knowledge-based inference," Connection Science 1(3), 1989, 327-342.

[MULL01]   Mullis, K. B., "The Unusual Origin of the Polymerase Chain Reaction," Scientific American, April, 1990, 56-65.

[YOON01]   Yoon, Y. O. , J. Neural Network Computing, 1(1) 1989.

A NEURAL NETWORK FOR CONTOUR POINT DETECTION

Gene A. Tagliarini and Edward W. Page

Department of Computer Science
Clemson University
Clemson, South Carolina 29634–1906

ABSTRACT

In order to analyze an image it is important to be able to identify and classify the objects that are present in it. Humans rapidly combine a variety of types of information such as object boundaries, shading, disparity and curvature to determine the classification of an object. Boundary contours are particularly useful for identifying an object with both human and machine vision systems. Numerous algorithms for automatically identifying the edges in an image have been explored and a review of these methods may be found in [1]. In the following, an artificial neural network (ANN) implementation of a boundary contour point detector is described.

As part of a continuing research effort, we designed an ANN for use in an object recognition system [7,8]. The input for this network consisted of features that were extracted from a digitized image. The outline of each observed object and its vertices were determined. This information was used to categorize each vertex on the basis of its interior angle. The vertex categories and the distances between the vertices formed the input to the ANN which in turn found the most compatible match of the observed object's features to the features of objects previously stored. On the basis of the feature match, the object was categorized. While the feature matching network employed neural network concepts, feature extraction was done conventionally. Subsequent study revealed that the boundary contour points could be located (in parallel) by an appropriately prescribed ANN whose neurons performed perceptron–like input/output functions and which were connected in an "on–center–off–surround" [2,3] pattern.

The ANN that was designed employs perceptrons to detect boundary contour points in continuously shaded images. The input/output relationship for a perceptron is a binary threshold function of the weighted sum of its inputs. Such neurons may be combined as illustrated in Fig. 1 to produce a cell assembly whose output would be 1 outside a small range of input values and 0 within that range. The combined effect of the assembly is the notch shaped output illustrated at the right which approximates the Mexican hat operator [4,5,6]. These cell assemblies are used to detect intensity gradients in the image.

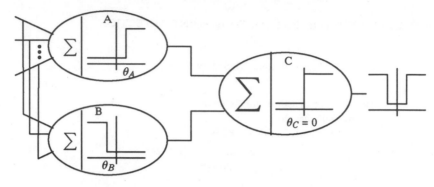

Fig. 1. A Cell Assembly

Figure 2 illustrates the connections that are required by the network. The input pixel array represents the digitized image. Four of the cell assemblies discussed above and one additional output perceptron are used for each pixel to be examined. The output perceptron indicates if the corresponding input pixel is at the boundary of a region by monitoring the response of the gradient–detecting cell assemblies. While the design shown here exploits only information from the horizontal and vertical near neighbors, it may be extended to detect other intensity gradients, such as those in the remaining near neighbors, as well.

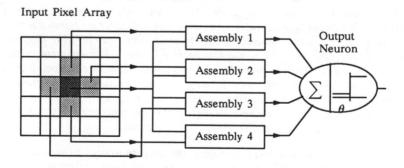

Fig. 2. Interconnections for a Single Pixel

A review of the recent results and further details of the architecture will be presented in the talk.

References

1. Chen, J. S., and G. Medioni, "Detection, Localization, and Estimation of Edges," IEEE Transactions on Pattern Analysis and Machine Intelligence, Vol. PAMI-11, No. 2, pp. 191–198, February 1989.

2. Grossberg, S. and E. Mingolla, "Neural Dynamics of Perceptual Grouping: Textures, Boundaries, and Emergent Segmentations," Perception & Psychophysics, Vol. 38, No. 2, pp.141–171, 1985.

3. Grossberg, S., "Nonlinear Neural Networks: Principles, Mechanisms, and Architectures," Neural Networks, Vol. 1, No. 1, pp. 17–61, 1988.

4. Hartmann, G., "Recursive Features of Circular Receptive Fields," Biological Cybernetics, Vol. 43, pp. 199–208, 1982.

5. Hartmann, G., "Recognition of Hierarchically Encoded Images by Technical and Biological Systems," Biological Cybernetics, Vol. 57, pp. 73–84, 1987.

6. Marr, D., and E. Hildreth, "Theory of Edge Detection," Proceedings of the Royal Society of London, B, Vol. 207, pp. 187–217, 1980.

7. Xie, Y., G. A. Tagliarini and E. W. Page, "Feature Matching Using Neural Networks," Neural Networks: Opportunities and Applications in Manufacturing, Society of Manufacturing Engineers, Detroit, MI, April 3–4, 1990.

8. Tagliarini, G. A. and E. W. Page, "Solving Constraint Satisfaction Problems With Neural Networks," Proc. IEEE First Int'l. Conf. on Neural Networks, Vol. 3, pp. 741–747, San Diego, CA, June 1987.

PULSE TRANSMISSION NEURAL NETWORKS:

TEMPORAL CONSIDERATIONS

Judith E. Dayhoff

Judith Dayhoff & Associates, Inc.
11141 Georgia Ave., Suite 206
Wheaton, MD 20902
301-933-9000

ABSTRACT

In Pulse Transmission (PT) Neural Networks,
processing units send pulses to one another, as do
most biological neurons. Here we introduce pulse
transmission (PT) neural networks as a class of
artificial neural network. We describe a PT
network model and show how it is inspired by
biological neural systems. We identify structures
that are important in the temporal dynamics and
capabilities of these networks, and show
similarities with recurrent neural networks.

INTRODUCTION

Pulse Transmission (PT) Neural Networks are
a class of neural network architecture in which
the processing units generate and receive pulses.
Most biological nerve networks can be considered
a type of pulse transmission network, as a
simplified model, because biological neurons
usually communicate by sending action potentials
from one neuron to another. These action
potentials are considered here to be pulses, or

265

impulses. We **show** a simplified biological model
that motivates our engineering approach to
building pulse transmission (PT) neural networks.
The internal dynamics and temporal processing of
the resulting PT models appear to create
significant capabilities in the temporal domain.

Figure 1 shows a PT neural network with three
layers - an input layer, a hidden layer, and an
output layer. Pulses are shown arriving as inputs
to the network, moving internally within the
network, and emerging as outputs from the network.
Inputs and outputs of the PT network are
spaciotemporal patterns of pulses (i.e.,
simultaneous pulse trains). Transmissions between
layers are also spaciotemporal patterns of pulses.

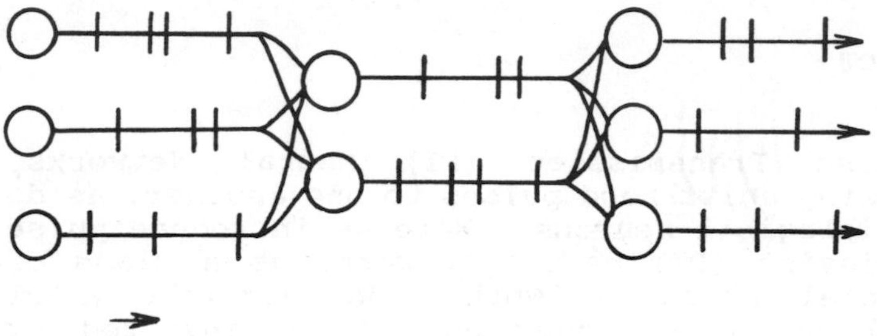

Figure 1

A PT neural network must have a mechanism at
each node for processing incoming pulses, and a
criterion for generating outgoing pulses. In this
paper, we describe these mechanisms for a PT
neural network paradigm.

A SIMPLIFIED BIOLOGICAL MODEL

Biological neurons have many complex details
to their anatomy and physiology that are not yet
in today's simulation models or in artificial
neural network paradigms. We have developed here

a simplified model of biological neurons, that
uses only a basic scheme of how pulses are
generated by each neuron, and how neurons
integrate arriving pulses. In this model, certain
parameters and structures influence the temporal
integration done by the neurons when processing
incoming pulses.

Figure 2 shows a biological neuron, as a
schematic drawing. Action potentials (pulses) are
usually generated at the spike initiation zone,
shown below the cell body. Pulses propagate down
the axon to the terminal points, where they
initiate biochemical changes at synapses. These
changes influence other (target) cells in the
network.

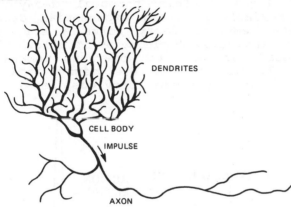

Figure 2. (Reprinted from [1].)

The top of Figure 2 shows the dendritic tree,
the major input area for the biological neuron.
Incoming pulses usually arrive from other axons at
synapses to the dendrites. In a biological
neuron, the dendritic tree contributes to a complex
mechanism of summing and weighting incoming
pulses. Here we show a simplified scheme that
reflects the basic dynamics of the membrane's
summing mechanism.

The cell membrane has a potential difference between the inside and the outside, due to unequal concentrations of charged particles on the two sides of the membrane. Incoming pulses cause changes in this membrane potential; when the membrane potential reaches a threshold value, an impulse is generated at the spike initiation zone.

Figure 3 illustrates a simplified description of the biological neuron's integration process. Initially the membrane potential is at a resting level. Incoming pulses cause a sudden increase (excitation) or a sudden decrease (inhibition).

When new pulses are not arriving, a temporal decay moves the membrane potential slowly towards its resting level. This temporal decay is dictated approximately by an exponential decay process. The temporal decay forms a sort of memory for the individual neuron, as it allows the neuron to retain in part its previous membrane potential level. Slowly this level decays back to the resting level, and the previous level is "forgotten".

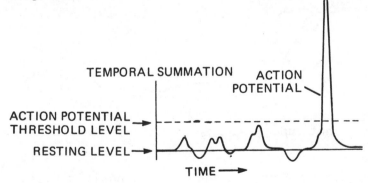

Figure 3. (Reprinted from [1].)

The neuron has a threshold level, also shown in Figure 3. When incoming pulses cause the membrane potential to reach threshold, the neuron fires a pulse. The pulse moves down the axon to target neurons. A higher threshold level tends to give the neuron more time to integrate incoming pulses, and a lower threshold allows the neuron to fire more often, after receiving fewer incoming pulses.

ARTIFICIAL PULSE TRANSMISSION (PT) NETWORK

Our PT network follows the model given above. Figure 1 shows an example PT network and Figure 3 illustrates the basic model of the PT processing unit. Equations govern the temporal integration, the temporal decay, and the criterion for generating a pulse.   The temporal decay is governed by the following equation:

(1) $a_j(t) = a_j(t_0) \, e^{-rt}$

where t_0 is the time of the most recent pulse that has arrived, t is the time elapsed since time t_0, $a_j(t_0)$ is the activity level of unit j at time t_0, and r is the decay constant for the exponential decay.

The summation of pulses is governed by an equation that updates the activity level by a fixed increment or decrement. If a pulse arrives at time t, then the activity level of unit j is updated as follows.

(2) $a_j(t) \longleftarrow a_j(t) + c$

where t is the time of pulse arrival, c is the amount of increment (positive or negative), j is the receiving processing unit, and $a_j(t)$ is the activity level of unit j at time t.

The generation of new, outgoing pulses is governed by the following equation:

(3)   If $a_j(t) > T$ then the update is

$$a_j(t) <-- R$$

where

T = threshold level

R = resting level

IMPULSES AND TEMPORAL INTEGRATION

There appears to be a relationship between the use of impulses by a PT network and the performance of temporal integration at each processing unit. The simplified biological model illustrates such a combination: when pulses are received, the processing unit has a means of integrating those pulses over time. In our models, the presence of a single pulse by itself is not significant; many pulses are needed for a neuron to cross threshold and fire. These pulses arrive over a period of time from one or more incoming lines.

We identify the following components that have temporal properties of the PT network model:

1. Threshold-to-resting difference (D). Figure 4 illustrates two alternative threshold-to-resting differences. In the first case, D is large, and it takes a relatively large number of arriving pulses spaced closely in time to reach threshold. In the second case, D is small, and fewer pulses are needed to reach threshold. This difference

(D) tends to reflect the amount of time that the
unit integrates incoming pulses before it
generates a new pulse.

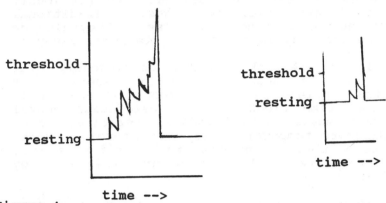

Figure 4

2. The decay constant. This governs how fast the
activity level decays over time. A larger decay
constant means a more rapid decay and thus a
shorter "memory". A smaller decay constant means
a slower decay and a longer "memory".

3. The relative length of inhibitory and
excitatory influences. These lengths are governed
by the exponential decay constant. A longer
excitation time (i.e., a slower decay) allows a
longer influence for each pulse. It gives more
time to sum and allows near-coincidences to be
farther apart in time and still to be summed. A
longer decay time for inhibition blocks the
influence of arriving excitatory pulses for
longer, which is especially important when pulses
do not arrive very often.

RECURRENT NETWORKS

It is of value to compare the PT pulse transmission neural network here to previously described artificial paradigms. Traditional neural networks do not use pulses. They also do not use temporal integration or temporal decay.

The PT network operates continuously over time, with pulses arriving or generated at arbitrary points in time, without the use of an internal time clock. Other artificial neural networks usually perform updates iteratively, done at discrete temporal steps. PT neural networks update their processing unit states asynchronously, and updates are triggered by arriving pulses.

We have identified the recurrent network as a paradigm that is closer to the PT network model than most other artificial architectures. Recurrent networks were described in [2]. These networks can have feedback loops that are comprised of an interconnection line from a processing unit going back to itself. Other feedback loops are also possible in recurrent networks. Recurrent networks can be configured to have internal temporal structure and to perform pattern-mapping tasks in the temporal domain.

Here we select recurrent networks for comparison with the PT network model described above. Recurrent networks are usually layered. Figure 5 shows processing units that send interconnection lines to themselves. A weight is associated with this interconnection. Activity levels for the hidden layer in Figure 5 are updated as follows.

(4)
$$a_j(t) = v_j a_j(t-1) + \sum_i w_{ji} a_i(t)$$

where the sum is taken over all units i in the layer below unit j. The variable v_j stands for the weight on the feedback loop of unit j, and w_{ji} is the weight to unit j in the hidden layer from unit i in the input layer.

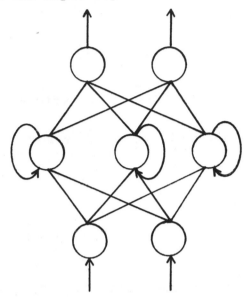

Figure 5

The first term in Equation (4) amounts to a time-stepped exponential decay. Here we assume that the weight v_j is fixed over time, and is between 0 and 1. During each time step, the fraction v_j is multiplied by the activity level. The remaining level is $v_j a_j(t-1)$.

In exponential decay, the amount of decay is fixed for each equal time interval. Thus, each time step results in the same percentage decrease. From Equation (1), after a time interval s, a fraction e^{-rs} of the original activity level is retained. Set v_j equal to that fraction, and let $a_j(t-1)$ be the initial activity level. Then, after a time interval s, the amount remaining from the exponential decay is $v_j a_j(t-1)$, the first term in (4). Thus the PT network's temporal decay can be approximated by a recurrent network's feedback loops.

SIMULATION MODELING

We have implemented a PT network that simulates the model described in this paper. Equations 1-3 are followed in this simulation model. Temporal properties were examined through simulation.

The resulting networks could perform pattern-mapping in which a set of input spactiotemporal (pulsed) patterns were mapped to a set of output spaciotemporal (pulsed) patterns. They could integrate over time and still perform pattern-mapping. They can allow time stretching and contracting of input patterns, and still generate the same or similar output patterns. The networks are also resilient to temporal gaps in the incoming data.

We have experimented with a variety of values for threshold-to-resting levels and temporal decay rates. Although the patterns used in these experiments were simple, we have found temporal integration abilities and resilience to temporal noise for the overall network.

REFERENCES

1. J. E. Dayhoff, 1990. _Neural Network Architectures: An Introduction_. New York: Van Nostrand Reinhold.

2. D. E. Rumelhart, J. L. McClelland, and the PDP Research Group, 1986. _Parallel Distributed Processing_, Vol. 1. Cambridge, Massachusetts: MIT Press.

APPLICATION OF FEEDBACK NETWORKS TO THE SOLUTION OF MASSIVE MARKOV PROCESSES

Robert Geist   Robert Reynolds   Darrell Suggs

Department of Computer Science, Clemson University
Clemson, South Carolina 29634-1906

INTRODUCTION

We are interested in a class of Markov processes that arises naturally in attempting to quantify the aesthetics of *digital halftone resolutions*. The digital halftone resolution problem may be stated as follows: given an $n \times n$ array V of real numbers, $V_{i,j} \in [0,1]$, produce an $n \times n$ array ω of binary integers, $\omega_{i,j} \in \{0,1\}$, such that ω, when displayed on a binary output device such as a computer monitor or laser printer, is a "good" representation of the real information, the intensities contained in V. The obvious resolution algorithm, round the values in V, fails to satisfy most interpretations of "good". For instance, if $V_{i,j} = .4999999$ for all i, j, then $\omega_{i,j}=0$ for all i, j, and a desired gray image is displayed as white. Consideration of neighborhood intensities seems imperative.

The most commonly used resolution algorithm is probably the *ordered dither* [1], in which we tile the image matrix V with a smaller fixed array D of threshold values, and then turn on the pixel (set $\omega_{i,j}=1$) if and only if $V_{i,j}$ exceeds the corresponding threshold value. A standard 4×4 tile is

$$D = \begin{bmatrix} 1/32 & 17/32 & 5/32 & 21/32 \\ 25/32 & 9/32 & 29/32 & 13/32 \\ 7/32 & 23/32 & 3/32 & 19/32 \\ 31/32 & 15/32 & 27/32 & 11/32 \end{bmatrix}$$

Note that a uniform intensity of 0.5 would cause 8 of every 16 pixels (every other one) to be turned on. Also note that the entries in the tile are carefully chosen to break horizontal and vertical lines, which are easily recognized by the eye.

In figure 1 we show a 256×256 pixel image resolved by this ordered dither. This image was printed on a conventional 300 pixel per inch laser printer at an expanded resolution of 75 pixels per inch.

Although the ordered dither algorithm can be highly parallel in implementation, one of the standard complaints lodged against this technique is that it imparts an artificial texture to the image. This "computery" look is quite evident in our figure.

Many such halftone resolution algorithms have been proposed (see [6]). One reason for the multitude of algorithms is that the usual measure of success for halftoning, the quality of the output image, has been gauged in a largely subjective manner.

275

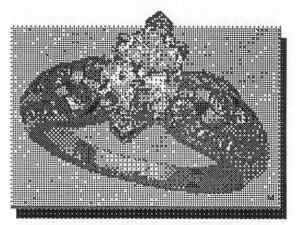

Figure 1: Resolution by ordered dither.

In [2] and [3] we took some initial steps towards a quantification of the quality of halftone images. We proposed a maximum-entropy Gibbs measure, P, for assigning numerical values to resolutions and a Hopfield neural network for selecting a resolution ω with relatively high value, $P(\omega)$. We also noted a natural connection with certain continuous-time Markov process.

In this paper we wish to consider an alternative, discrete-time Markov process whose steady-state (invariant measure) is precisely our maximum-entropy Gibbs measure. We consider a parallel simulation of this Markov process as an alternative technique for selecting a resolution ω with high value, $P(\omega)$. We compare this approach with the previously suggested Hopfield neural network approach in terms of expected image quality. Sample halftone resolutions using both techniques are provided.

A GIBBS MEASURE

Hereafter we number pixels in our $n \times n$ array left-to-right and top-to-bottom using the non-negative integers and refer to intensities $\{V_i \mid i = 0, 1, 2, ..., N - 1\}$.

Consider a formulation of the halftone resolution problem in which we view the set of pixels, S, as a set of vertices of a connected graph. Edges exist from each vertex (pixel) to all vertices in a physical neighborhood of some fixed radius.

Regard the set $T = \{\text{pixel off, pixel on}\} = \{0,1\}$ as a set of labels to be attached to the elements of S. If we represent the power set of S by the cartesian product, $\{0, 1\}^{|S|}$, then a *potential* on S is a map

$$U : \{0, 1\}^{|S|} \times T^{|S|} \to R$$

such that

1. $U(\emptyset, \omega) = 0$

2. $U(A, \omega) = U(A, \omega')$ whenever label vectors ω and ω' agree on subset A.

The *energy* of the potential is then given by

$$U(\omega) = \sum_A U(A, \omega),$$

and the *Gibbs measure induced by U* [7] is the probability measure induced on the collection of possible labelings, $T^{|S|}$, given by

$$P(\omega) = \frac{e^{-U(\omega)}}{M},$$

where M is a normalizing constant.

The chief motivation for consideration of such measures is their relationship to entropy. Among all measures with the same expected energy, $E[U(\omega)] = \sum_\omega P(\omega) U(\omega)$, it is the one that maximizes the entropy, $- \sum_\omega P(\omega) log P(\omega)$.

We now specify a potential U on our pixel graph as follows: for each single-element set $\{i\} \subset S$, let

$$U(\{i\}, \omega) = (1 - 2\omega_i)(2V_i - 1), \tag{1}$$

where $V_i \in [0, 1]$ is desired pixel intensity. Note that as V_i becomes larger, energy reduction mandates $\omega_i = 1$ (on). For each two-element set $\{i, j\} \subset S$ where i and j are neighbors, let

$$U(\{i, j\}, \omega) = (1 - 2\omega_i)(1 - 2\omega_j) f(i, j). \tag{2}$$

where $f(i, j)$ is a symmetric function that represents the force with which we insist that neighbor pixels assume opposite parity. For all other sets $A \subset S$, let $U(A, \omega) = 0$.

Specification of $f(i, j)$ is a delicate matter. If we regard the value of pixel i, ω_i as a stationary stochastic process with mean μ, then its *autocovariance sequence* (in space) is given by

$$R_k = E[(\omega_i - \mu)(\omega_{i+k} - \mu)] \quad k = 0, 1, \ldots$$

where $R_0 = \sigma^2$, the variance. The associated *autocorrelation sequence* is given by

$$\rho_k = \frac{R_k}{R_0} \quad k = 0, 1, \ldots.$$

and the *power spectrum* is the Fourier transform,

$$P(\lambda) = R_0 + 2 \sum_{k=1}^{+\infty} R_k cos(k\pi\lambda) \quad 0 \leq \lambda \leq 1$$

Note that zero spatial correlation of pixel value would give $R_k = 0, k > 0$ and hence $P(\lambda) = \sigma^2$, all λ. A constant frequency function such as this is often regarded as white noise.

Ulichney argues [11] that for stochastic halftone resolution of a fixed intensity, $V_i = \mu$, all i, it is the low frequency noise that gives rise to unpleasant visual effects. If we remove low frequency noise from the white noise function, $P(\lambda) = \sigma^2$, we obtain

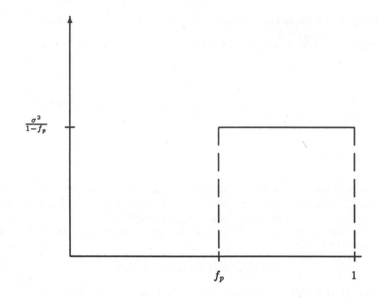

Figure 2: Blue noise.

the "blue noise" spectrum of figure 2. Here f_p is the *principal frequency* which is taken from the desired constant intensity μ:

$$f_p = \left\{ \begin{array}{ll} \sqrt{\mu} & 0 \leq \mu \leq 1/2 \\ \sqrt{1-\mu} & 1/2 < \mu \leq 1 \end{array} \right.$$

Inverting this idealized spectrum, we obtain

$$\begin{array}{rcl} R_k & = & \int_0^1 P(\lambda) cos(k\pi\lambda) d\lambda \\ & = & \dfrac{\sigma^2}{1-f_p} \left[\dfrac{-sin(k\pi f_p)}{k\pi} \right] \end{array}$$

and so

$$\rho_k = \frac{-sin(k\pi f_p)}{(1-f_p)k\pi}$$

We can take this correlation coefficient ρ_k as the desired relative strength of connection between two pixels at distance k in a halftone resolution of a region of fixed intensity value μ.

Thus our connection structure is specified as follows: for each pixel i let μ_i denote the mean of the intensities, the V's, in a neighborhood of radius R about pixel i, and let $\mu_{i,j} = (\mu_i + \mu_j)/2$. If pixel j is at distance $k \leq R$ from i, we set

$$f(i,j) = \frac{Csin(k\pi\sqrt{\mu_{i,j}})}{(1-\sqrt{\mu_{i,j}})k\pi}.$$

where $C > 0$ is a scale factor. For i and j not neighbors, let $f(i, j) = 0$.

If $U(\omega) = \sum_{A \subset S} U(A, \omega)$ is the energy of this potential, then the Gibbs measure induced by U,

$$P(\omega) = \frac{e^{-U(\omega)}}{M},$$

is a natural assignment of measure to the set of all possible halftone resolutions of the given set of intensities, $\{V_i \mid i=0,1,...,N\text{-}1\}$. Our task is thus reduced to selecting that ω with maximum $P(\omega)$ or, equivalently, with minimum $U(\omega)$. Unfortunately, even for a relatively small image of size 256×256 pixels, the set of all resolutions is of size $2^{2^{16}} \approx 10^{20,000}$, which dwarfs even the number of atoms in the universe. Selection of that ω with minimum energy thus requires a special mechanism.

A NEURAL NET

One such mechanism is the class of neural networks proposed by Hopfield [5]. A four element example, from [10], is shown in figure 3. Neurons are represented by

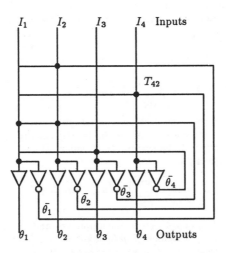

Figure 3: Four element Hopfield network.

amplifiers, each providing both standard and inverted outputs (voltage $\theta_i \in [-1, 1]$). Synapses are represented by the physical connections between input lines to the amplifiers and, in feedback, output lines from the amplifiers. Resistors are used to make these connections. If the input to amplifier i is connected to the output of amplifier j by a resistor with value R_{ij}, then the conductance of the connection is T_{ij}, whose magnitude is $1/R_{ij}$ and whose sign is determined by whether the connection to amplifier j is from the standard or inverted output. Hopfield showed that when the matrix T is symmetric with zero diagonal and the amplifiers are operated in "high-gain" mode, the stable states of the network are binary ($\{-1,1\}$) and are the local minima of the computational energy,

$$E(\theta) = (-1/2) \sum_{i=0}^{N-1} \sum_{j=0}^{N-1} T_{ij}\theta_i\theta_j - \sum_{i=0}^{N-1} \theta_i I_i. \tag{3}$$

Here I_i is the external input to amplifier i.

We observe that our energy function $U(\omega)$ can be written in the form of the Hopfield computational energy. Specifically, if we assign

$$T_{ij} = T_{ji} = \begin{cases} -f(i,j) & i, j \text{ neighbors} \\ 0 & \text{otherwise} \end{cases}$$

$$I_i = (2V_i - 1)$$

$$\theta_i = (2\omega_i - 1)$$

in the Hopfield computational energy expression (3), we obtain $U(\omega)$.

We should note here that any selection of connection parameters is likely to benefit from consideration of local image intensity. Specifically, a reasonable constraint on resolution ω is that $\sum_{i \in nhbd} \omega_i$ should not deviate greatly from $\sum_{i \in nhbd} V_i$.

To incorporate the additional constraint, we use a variation on a technique suggested in [10]: let $m = \lfloor \sum_i V_i + 0.5 \rfloor$, the rounded total intensity. A local energy of the form

$$E_{nhbd}(\theta) = (\sum_{i \in nhbd} \frac{\theta_i + 1}{2} - m)^2 + \sum_{i \in nhbd} \frac{1 - \theta_i^2}{4}$$

is then within the Hopfield framework and would tend to force m neurons on.

Net simulation to obtain a local minimum is traditionally approached (e.g. [10]) as a numerical integration of the system of N differential equations describing the operation of the amplifiers [5]:

$$C_i du_i/dt = \sum_j T_{ij}g(u_j) - u_i/R_i + I_i. \tag{4}$$

Here the u_i are internal input voltages to the amplifiers, and are related to the desired output voltages, the θ_i, by a sigmoidal *gain function*, $g(x)$. A reasonable choice for $g(x)$ is a scaled hyperbolic tangent, $g(x) = tanh(\lambda x)$. Here λ is called the *gain*. The C_i are the input capacitances of the amplifiers, and $R_i = 1/(1/\rho + \sum_j |T_{i,j}|)$, where ρ is amplifier input resistance.

We have found numerical integration of large (2^{16} neuron) systems of the form (4) to be extremely time consuming, and therefore have developed an alternative approach. Any equilibrium of (4) is given by

$$0 = \sum_j T_{i,j}g(u_j) - u_i/R_i + I_i \tag{5}$$

that is,

$$u_i = R_i(\sum_j T_{i,j}g(u_j) + I_i)$$

or simply

$$u = G(u),$$

where $G(u) = diag(R)(Tg(u)+I)$, $diag(R)$ has R_i's on the diagonal and 0's elsewhere, and $g(u) = (g(u_1), g(u_2), ...)$. Thus we seek a fixed point of a certain N-dimensional function. A numerical iteration scheme [2] is used to solve for the fixed point.

In figure 4 we see the results of applying our neural net solution scheme to

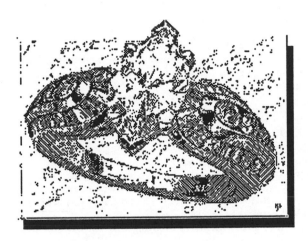

Figure 4: Resolution by neural net.

the same data used in figure 1. Marbling in the background has begun to emerge. Greater detail is visible in the ring, including a signature initial in the lower right and a suggestion of initials on the inside of the ring band.

A MARKOV MODEL

Consider again the set of all possible resolutions, ω, as the state space of a Markov process. Transitions are allowed from any resolution ω to any resolution $\omega^{\delta i}$ that differs from ω in exactly one pixel, i. The transition rates are given by:

$$Q(\omega, \omega^{\delta i}) = q(i)e^{min\{0, U(\omega)-U(\omega^{\delta i})\}}$$

where $q(i) > 0$ and $\sum_i q(i) = 1$. The key observation about this process is

$$
\begin{aligned}
P(\omega)Q(\omega, \omega^{\delta i}) &= q(i)e^{min\{0, U(\omega)-U(\omega^{\delta i})\}}e^{-U(\omega)}/M \\
&= q(i)e^{min\{-U(\omega), -U(\omega^{\delta i})\}}/M \\
&= P(\omega^{\delta i})Q(\omega^{\delta i}, \omega)
\end{aligned}
$$

which is precisely the condition needed to conclude that our Gibbs measure, $P(\omega) = e^{-U(\omega)}/M$, is the steady-state (invariant measure) of the process [8].

Straightforward simulation of this Markov process (select pixel i with probability $q(i)$; then change i with probability $e^{min\{0, U(\omega) - U(\omega^{\delta i})\}}$) is thus an alternative approach to selecting a resolution ω with high $P(\omega)$. If we start with a random resolution, then after many iterations we should achieve resolution ω with probability $P(\omega)$. Desirable resolutions are thus more likely to appear than undesirable ones.

When $q(i) = 1/N$ this type of Markov simulation has been called simulated annealing [9] or the Boltzmann machine [4]. A more reasonable choice of $q(i)$ in our context would be

$$q(i) = \frac{|2V_i - 1|}{\sum_i |2V_i - 1|},$$

thus ensuring that high and low intensity pixels are resolved properly.

We should also note that a completely parallel implementation here is straightforward. From our expressions for energy, (1) and (2), we see that if i and j are not neighbors then

$$U(\omega) - U(\omega^{\delta i}) = U(\omega^{\delta j}) - U(\omega^{\delta j \delta i})$$

Thus when pixel i is selected for update, we need only prevent modification of the neighbors of i. Other pixels may be updated simultaneously.

Convergence can clearly be hastened by adding a scale factor $K > 1$ to energy U. An increase in K corresponds to the notion of "cooling" in simulated annealing. Note that if we use the equivalence $U(\omega) = E(\theta)$ to translate this scaling factor into multipliers of $T_{i,j}$ and I_i, then from (5) we see that an equivalent system is obtained by a scaling of resistance, $\rho \rightarrow \rho K$. Thus the notion of cooling could be introduced into the Hopfield network approach as well.

In figure 5 we see the results of applying this Markov simulation to the same data.

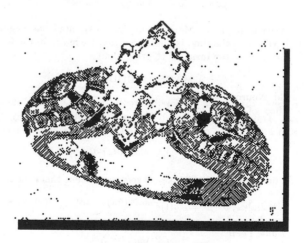

Figure 5: Resolution by Markov simulation.

We see that the amount of detail in the ring is similar to that of the neural net resolution, including the initials on the ring band. However, some ring edges have been lost and the marbling in the background is only hinted at. Several attempts to bring out the background marbling via a consideration of local intensity similar to that used in the neural net solution failed to do more than produce an undifferentiated gray background.

CONCLUSIONS

We have considered two techniques for finding highly likely states of extremely large Markov chains: neural networks and Boltzmann machines. When applied to the Markov process arising from a digital halftone resolution problem, both techniques produce pictures superior to that resulting from the standard ordered dither. The neural net resolution contains considerably more information than that of the Boltzmann machine, probably because the neural net admits easily to a modification which considers local intensity. Further investigation into incorporating local intensity into the Boltzmann machine is obviously warranted.

REFERENCES

[1] J. Foley and A. Van Dam. *Fundamentals of Interactive Computer Graphics.* Addison-Wesley, Reading, Massachusetts, 1984.

[2] R. Geist and R. Reynolds. Colored noise inversion in digital halftoning. *Proc. of Graphics Interface '90 (GI '90)*, pages 31–38, Halifax, Nova Scotia, May, 1990.

[3] R. Geist and R. Reynolds. The most likely steady state for large numbers of stochastic traveling salesmen. *Proc. First Int. Workshop on the Numerical Solution of Markov Chains*, pages 569–581, Raleigh, NC, January, 1990.

[4] G. Hinton, T. Sejnowski, and D. Ackley. Boltzmann machines: Constraint satisfaction networks that learn. *Tech. Rep. CMU-CS-84-119*, Dept. of Comp. Sci., Carnegie Mellon Univ., 1984.

[5] J.J. Hopfield. Neurons with graded response have collective computational properties like those of two-state neurons. *Proc. Natl. Acad. Sci.*, **81**:3088–3092, 1984.

[6] J. Jarvis, C. Judice, and W. Ninke. A survey of techniques for the display of continuous tone pictures on bilevel displays. *Comp. Graphics Image Process.*, 5:13–40, 1976.

[7] J. Kemeny, J. Snell, and A. Knapp. *Denumerable Markov Chains.* Springer-Verlag, New York, second edition, 1976.

[8] R. Kindermann and J Snell. *Contemporary Mathematics: Vol. 1, Markov Random Fields and Their Applications.* AMS, Providence, RI, 1980.

[9] S. Kirkpatrick, C. Gelatt, and M. Vecchi. Optimization by simulated annealing. *Science*, **220**, 1983.

[10] E. Page and G. Tagliarini. Algorithm development for neural networks. In *Proc. SPIE Symp. on Innovative Science and Technology*, Los Angeles, CA, January 1988.

[11] R. Ulichney. *Digital Halftoning*. MIT Press, Cambridge, Massachusetts, 1988.

LANGUAGE ENGINEERING

Pradip   Peter   Dey

Computer Science Department
Hampton University
Hampton, VA 23668

ABSTRACT

The main goal of language engineering is to
develop natural language systems, including human-
machine interfaces, incrementally. A   formalism
called   Tree Adjoining Grammar (TAG) and some
computer aided specification techniques   facilitate
incremental specification of a language.   Such
specifications coupled with   massive   preprocessors
will allow high quality human-machine interfaces in
medical and other application areas.

1.   INTRODUCTION

Usability of today's complex machines depends
on human-machine interfaces. Such interfaces play
an ever increasing role in computer based medical
systems. Language engineering is a new area of
research which is developing techniques that will
provide high quality human-machine interfaces in
natural language. Most natural language systems
include a syntactic and a semantic component. This
paper is mainly concerned with the   syntactic
component, which is primarily a parser. It attempts
to show that the syntactic component, i.e. the
parser,   can be developed incrementally using
language engineering techniques. It can be assumed
that semantics is syntax directed and can be
developed incrementally.

285

One of the difficulties in designing a parser has been in defining a subset of a natural language such as English according to its sentence length. We use a formalism called Tree Adjoining Grammar (TAG) to specify this subset. Parsing is viewed as a relation between strings and trees in TAG. One of the advantages of using this formalism is that it allows computer generated specification of such a subset. Furthermore, it allows computer generated specification to be mixed with hand-crafted specification. These ideas are briefly explained in section 2.2. The parser has an inefficient module that pregenerates all possible parse trees up to some length, where length of a parse tree is the number of leaves it has. However, this inefficient module is shifted in a pre-processor in order to achieve runtime efficiency.

2.   PARSING AS A RELATION BETWEEN TREES AND STRINGS

2.1   Tree Adjoining Grammar

Parsing is always based on a finite description of the language called grammar. TAG seems to be appropriate for our purposes because: (1) it can be easily decomposed into independent modules which can be processed concurrently, (2) it can be developed incrementally since it represents information about the language in a highly modular fashion, (3) it can be used for computer generation of syntactic structures that can be mixed with hand-crafted structures.

We give an introduction to an extended form of TAG in this section. For descriptions of standard TAG see [Joshi 1985, Vijay-Shankar and Joshi 1985, Dey and Hayashi 1990, Dey, Bryant and Takaoka 1990]. TAG defines a finite set of elementary trees and an adjunction operation that produces composite trees through the combination of simple (elementary) ones. The trees are taken to be structural descriptions of sentences (or clauses) in the language. The central module of the parser is a pattern matcher that finds every tree that matches the input string. If there is only one tree that matches the input then the input is structurally unambiguous. If the input matches more than one tree, it is structurally ambiguous. If there is no exact match then the best available approximate match is determined according to some heuristics. It is this later match that allows parsing of semi-grammatical sentences. A set of structural descriptions of simple sentences is stored into a tree-bank. The adjunction operation represents a derivation with structural descriptions. The grammar can be built incrementally simply by adding new structural descriptions of sentences into the tree-bank. The major components of the parser are shown in Figure 1.

(1)   Figure 1

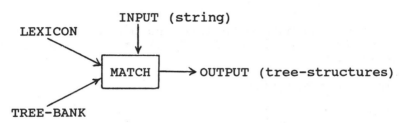

The tree-bank has two types of trees: initial trees and auxiliary trees. Auxiliary trees are used in a composition operation called adjoin to account for recursion; they do not occur independently in the language. We describe auxiliary trees with adjoining operation at the end of this section. You may assume for the time being that the tree-bank has only initial trees such as that in (3a) whose leaves (frontier nodes) are lexical category symbols. The lexicon is sorted in alphabetical order of keys to allow binary search.

(2)   LEXICON

```
(an      D)
(boy     N)
(flies   (V N))
(the     D)
(time    (N A V))
   :
   :
```

The tree-bank has initial trees like the one given in (3a). (N = Noun, V = Verb, D = Determiner, A = Adjective, P = Preposition, NP = Noun Phrase, VP = Verb Phrase, R = Relativizer, S = Sentence).

(3)a.   Tree1 =

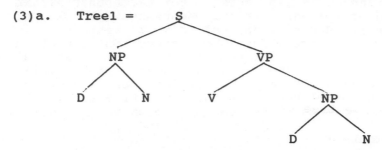

This tree matches a large number of sentences in English. Specifically, Tree1 matches sentences like (3b).

(3)b.   The boy likes the girl

Tree1 =

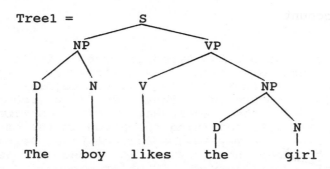

In the match phase, the leaves of the trees are matched against the categories of input words. Suppose that the input is the sentence given in (3b). There is a procedure in MATCH called 'CATEGORIES' that returns the lexical categories of the input string as shown in the list form below.

(CATEGORIES   '(The machine runs the program))

= (D N V D N)

When a lexical item belongs to two or more lexical categories it gives rise to lexical ambiguity which is a source of inefficiency in many parsers [Barton, Berwick and Ristad 1987]. Dey, Bryant and Takaoka [1990] show that lexical ambiguity can be processed efficiently in TAG. Parsing TAG is different from traditional parsing [Aho and Ullman 1972, Woods 1970 and Winograd 1983] and it has at least two advantages: (i) the trees are not built at run time, so it is fast, (ii) the trees of the tree-bank are grouped according to their length so that for a given input only a small part of the tree-bank has to be searched. Trees within each part of the tree-bank (called a sub-tree-bank) is ordered alphabetically so that a binary search can be executed. This reduces the search time to O(log x/y) where x is the number of trees, and y is the number of sub-tree-banks. By applying parallel binary search considerable speed-up can be achieved.

2.2 Recursive Adjoining

TAG accounts for natural language recursion by adjoining. An example of recursion is given in (4) where "who carries a radio" is an embedded clause.

(4) The boy who carries a radio likes the girl.

We mentioned earlier that the tree-bank has two types of trees: initial and auxiliary. An auxiliary tree is given in (5).

(5) Tree2 =

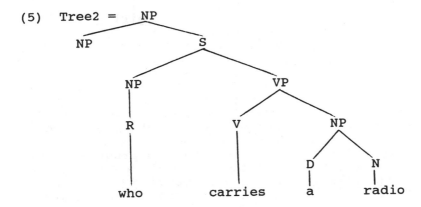

Unlike an initial tree, each auxiliary tree has a leaf which is identical to its root. This leaf is called a hook and plays an important role in adjoining. An auxiliary tree can be adjoined at any node of a non-auxiliary tree (initial or composite) if that node is labeled by the same symbol as the root of the auxiliary. An example of adjoining is shown in (6) in which Tree2 is adjoined at an NP node in Tree1.

(6)   Tree3 =

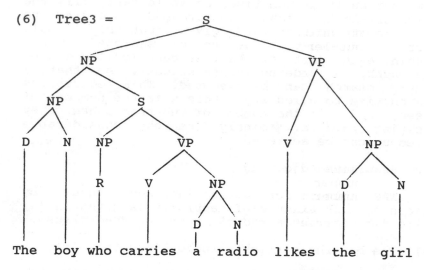

The boy who carries a radio likes the girl

If Tree2 is adjoined at the NP dominated by VP of Tree1 then we get sentences like (7). If Tree2 is adjoined at both the NPs of Tree1 then we get sentences like (8).

(7) The boy likes the girl who carries a radio.
(8) The boy who carries a radio likes the girl who drives a truck.

Notice that the match described earlier will work if it is preceded by adjoining. Adjoining is comparable to transformations of early transformational grammars [Chomsky 1957]. However, adjoin is applied before runtime to produce all possible trees for all sentences up to some length, say 50. These trees are then grouped according to their length and each group is stored in a separate sub-tree-bank. The length of a tree is the number

of leaves it has. Thus, the length of Tree3 given
in (6) is 9 and it is stored in sub-tree-bank-9.
The tree-bank is a sorted list of sub-tree-banks.
The leaves of each tree is pre-processed and stored
in association with the tree. Within each sub-
tree-bank the trees are stored according to
alphabetical order of their preprocessed leaves.
The runtime computation for parsing is limited to
(1) determining the length of the input, (2)
obtaining the lexical categories of each word of
the input by a binary search of the lexicon, (3)
determining which sub-tree-bank is relevant for the
given input-length, and (4) searching the relevant
sub-tree-bank by a binary search. How fast is this
parsing going to be? In order to discuss this
question the following parameters are used in the
expressions of the time complexity:

n: the length of the input string
l: the number of entries in the lexicon
k: the number of categories in the lexicon
i: the number of initial trees
m: the number of auxiliary trees
x: the number of non-auxiliary trees

It is easy to explain the combinatorial effect
of adjoining if we view parsing as a state-space
searching problem. Assume that the initial trees
represent a set of initial states. Adjoining of an
auxiliary tree to an initial or derived composite
tree represents an operator. The implicit search
space is defined by these operators and the initial
trees. Since there is no theoretical upper bound on
the number of recursive adjoining that can be
applied, the search space is potentially infinite.
However, if you know that people do not use more
than 4/5 recursive structures in a sentence then
you can assume that the depth of the search space
is about 5. Natural languages such as English have
some (practical) restrictions (such as center
embedding) on recursion. What is then a reasonable
estimate for x? Suppose that the branching factor
(degree) of each state is a constant b, and the
depth is d. Then, $x = b(b-1)i/(b-1)$, where d is
the number of recursive application of adjoin. $d <
n$; infact, d is usually a small fraction of n.
Based on these realistic assumptions, $x \leq c2^n$,

where c is a constant. The total running time for parsing TAG is   n log l + n + log x/n   in which log x/n is significant.   Since x is an exponential function of n the running time is in the order of O(n).

3.   FUTURE   PROSPECTS

In order to   ascertain   precisely   the   exact growth   rate   of   the   tree-bank,   adjoin has to be applied to exhaustive lists of English initial   and auxiliary   trees.   The   nature   of   semantics   of natural language is not yet fully understood and is often   a   matter   of   debate   [Woods 1968, McCawley 1968, Steinberg and Jakobovits 1971,   Jackendoff 1972,   Bobrow   and   Collins   1975, Keenan and Faltz 1985, Dowty,   Wall   and   Peters   1981,   Schank   and Birnbaum   1980].   It appears that a recent trend in this area is to use predicate calculus (logic)   to describe   meanings.   In   one of its simplest forms, the   meaning   is   given   by   translating   natural language   into   logic [Schubert and Pelletier 1982, Warren and Pereira 1982, Warren and Friedman 1982].

In   TAG   based   systems   context-independent semantic   structures   can   be generated in logic by associative   semantics.   Associated   with   each elementary   tree   in   the   tree-bank   is a semantic structure specifying a logical form.   For   certain types of ambiguity, two or more semantic structures would be associated with a single   syntactic   tree. If an elementary tree is brought into the syntactic derivation then the   predicate   and   terms   of   its associated   semantic   structure(s) are instantiated with the word meanings of the input sentence.   The problems   posed   by   ambiguity,   imprecision   and synonymy   have   to   be   carefully   studied.   The semantics   of   sentences matching against composite trees   are   developed   compositionally   from   the semantics   of initial and auxiliary trees.   A major problem to be investigated in   future   research   is the nature of semantic composition.

REFERENCES

Aho, A.V. and J. D. Ullman. 1972. The theory of parsing, translation and Compiling: volume 1. Englewood Cliffs, N.J.: Prentice-Hall.

Barton, G. E., Berwick, R., and Ristad, E. 1987. Computational Complexity and Natural Language, MIT Press.

Bobrow, D.G. and A. Collins 1975, (eds.) 1975. Representation and Understanding, New York: Academic Press.

Chomsky, N. 1957. Syntactic structures, Mouton.

Dey, P., and Hayashi, Y. 1990, "A multiprocessing model of natural language processing," Theoretical Linguistics, in print.

Dey, P., Bryant, B. and Takaoka, T. 1990, "Lexical ambiguity in tree adjoining grammars," Information Processing Letters, 34, 65-69.

Dowty, D. R., R. E. Wall and S. Peters 1981, Introduction to Montague Semantics, D. Reidel.

Jackendoff, R. S. 1972. Semantic Interpretation in generative grammar , MIT Press.

Joshi, A. K. 1985. "Tree Adjoining Grammars: How much context-sensitivity is required to provide reasonable structural descriptions?" Dowty, et al (eds.) Natural Language Parsing.

Keenan, E. L. and L. M. Faltz 1985, Boolean Semantics for Natural Language, Dordrecht: D. Reidel.

McCawley, J. D. 1968. "The role of semantics in a grammar". In E. Bach and R.T. Harms (eds), Universals in Linguistic Theory, New York: Holt, Rinehart and Winston.

Schank, R. and L. Birnbaum 1980. "Memory, Meaning, and Syntax," Report 189, Computer Science Dept. Yale University.

Schubert, L. and F. J. Pelletier 1982, "From English to Logic: Context-Free Computation of 'Conventional' Logical Translation," Computational Linguistics 8, 27-44.

Steinberg, D. and L. A. Jakobovits (eds) 1971, Semantics: An Interdisciplinary Reader in Philosophy, Linguistics and Psychology.

Stockwell, R. P., Schachter, P. and B. H. Partee 1973, The Major Syntactic Structures of English. New York: Holt, Rinehart and Winston.

Vijay-Shankar, K., and Joshi, A. K. 1985 "Some Computational Properties of Tree Adjoining Grammars", 23rd Annual Meeting of The Association for Computational Linguistics.

Warren, D. S. and J. Friedman 1982. "Using Semantics in Non-Context-Free Parsing of Montague Grammar," Computational Linguistics 8, 123-138.

Warren, H. D. and F. C. N. Pereira 1982, "An Efficient Easily Adaptable System for Interpreting Natural Language Queries," Computational Linguistics 8, 110-122.

Winograd, T. 1972. Understanding Natural Language, New York: Academic Press.

Winograd, T. 1983. Language as a cognitive process: Syntax. Reading, Mass.: Addison-Wesley.

Woods, W.A. 1970. "Transition Network Grammars for Natural Language Analysis". Communications of the ACM, 13, 591-606.

Woods, W.A. 1968. "Procedural Semantics for a question answering machine", Fall Joint Computer Conference, 457-471.

ACKNOWLEDGEMENT

The author is grateful to M. R. Ellis, A. K. Joshi, T. Takaoka, S. Iyengar, Y. Hayashi, B. R. Bryant, E. Battistella, K. Reilly, S. Madan and C. Wilson for their comments and/or help.

AUTHOR INDEX

-- A --

Anderson, L. R.   231

-- B --

Ballard, J. G.   147
Barineau, D. W.   51
Barry, P.   211
Benachenou, D.   251
Blichington, T. F.   5
Blum, J. J.   59

-- C --

Cader, M.   251
Cheatham, J. B.   211
Chen, K. C.   63
Chynoweth, D. P.   131
Clarke, A. M.   1,37
Conner, A. M. L.   1

-- D --

Dayhoff, J. E.   265
Denys, B. G.   155
Dey, P. P.   285
DiBianca, F. A.   175

-- E --

Endorf, R. J.   175

-- F --

Fan, N.   155
Ferreiro, J. I.   23
Foo, S. Y.   231
Fritsch, D. S.   175
Frey, E. C.   221
Frye, R. C.   241

-- G --

Geist, R. M.   275
Gerber, J. F.   137
Gisser, D. G.   201
Goble, J. C.   201
Grant, J. W.   81
Gullberg, G. T.   185,221

-- H --

Ha, B.   23
Hack, S. N.   165
Hale, S. A.   115
Harris, T. R.   33
Hendricks, J. L.   185
Henry, G. W.   23

-- I --

Ideker, R. E.   5
Isaacson, D.   201

-- J --

Johnston, R. E.   147

-- K --

Kachroo, A.   211
Kachroo, P.   211
Kaplan, D. T.   15
Kroushkop, T. A.   211

-- L --

Lee, D.   111
Lenhardt, M. L.   37
Li, C. C.   155,191
Liu, W.   175
Lucas, C. L.   23

295